# Understanding the Human Body: An Introduction to Anatomy and Physiology, 2nd Edition


Anthony Goodman, M.D., F.A.C.S.

**PUBLISHED BY:**

**THE GREAT COURSES**
**Corporate Headquarters**
**4840 Westfields Boulevard, Suite 500**
**Chantilly, Virginia 20151-2299**
**Phone: 1-800-832-2412**
**Fax: 703-378-3819**
**www.thegreatcourses.com**



Printed in the United States of America

## Anthony Goodman, M.D., F.A.C.S.

Montana State University


Dr. Goodman received his B.A. degree from Harvard College and his M.D. from Cornell Medical College. He trained as a surgical intern and resident at the University of Michigan Medical Center in Ann Arbor and completed his surgical training and chief residency at the Harvard Surgical Service of Boston City Hospital, New England Deaconess Hospital, Lahey Clinic, and Cambridge City Hospital.

Dr. Goodman served as surgeon on the hospital ship for Project H.O.P.E. and served with the U.S. Army Medical Corps from 1971 to 1973. From 1973 to 1992, he was a general surgeon, specializing in the surgical treatment of cancer, and was Clinical Associate Professor of Surgery at the University of Miami School of Medicine.

In 1991, Dr. Goodman was Visiting Professor of Surgery at the Christchurch Clinical School of Medicine in Christchurch, New Zealand. He has served as an examiner for the American Board of Surgery and is a Fellow of the American College of Surgeons and a Diplomate of the American Board of Surgery. He has been a member of the American Society of Colon and Rectal Surgeons and the British Association of Surgical Oncology, as well as founder of the Broward Surgical Society. He has published numerous papers on both clinical and experimental surgery.

At present, Dr. Goodman is Adjunct Professor in the Department of Microbiology, Montana State University, and Adjunct Professor in the W.W.A.M.I. Medical Sciences Program at Montana State University, where he teaches gross anatomy.

Dr. Goodman is also the author of a work of historical fiction, *The Shadow of God: A Novel of War and Faith* (Sourcebooks Landmark, Naperville, IL, 2002). ■

# Disclaimer

These lectures on Human Anatomy and Physiology are intended to increase the understanding of the structure and function of the human body. They are in no way designed to be used as medical references for the diagnosis or treatment of medical illnesses or trauma.

Neither The Teaching Company nor Dr. Goodman can be responsible for any result derived from the use of this material. Questions of diagnosis or treatment of medical conditions should be brought to the attention of qualified medical personnel.

# Table of Contents

## INTRODUCTION



## LECTURE GUIDES

## Table of Contents

# Table of Contents

## Table of Contents

# Understanding the Human Body: An Introduction to Anatomy and Physiology, 2nd Edition

## Scope:

This series of lectures will focus on the structure and function of the human body, its anatomy and physiology. The study of anatomy alone, without reference to both the normal and abnormal function of the human body, has little meaning. However, when studied in the context of the exquisite and intricate relationships of anatomy to those normal processes that keep us alive and allow us to reproduce and evolve, the subject becomes a gripping page-turner.

Human gross anatomy is the study of the structure that can be seen by the unaided eye. Microscopic anatomy, called *histology*, is the study of those structures too small to be seen without the help of a microscope. Together, they make up the study of the structure of the human body. Gross anatomy is the single most time-consuming course for the first-year medical student, who spends months in the laboratory dissecting an embalmed human cadaver. This right to dissect a human body was the result of hard-fought battles with both church and state, taking place over centuries. It is regarded by physicians and surgeons as one of the most important privileges in our medical education, and it is treated with the greatest respect. Disrespect of any kind for those who donated their bodies for our education is not tolerated.

In the dissection of a cadaver, anatomy is studied using *regional anatomy*. Organs are studied in one area at a time, and understanding their relationships to each other is extremely important. Indeed, it is of utmost importance that the surgeon is thoroughly knowledgeable about which organs lie directly next to, above, behind, and beneath each structure that he or she may cut through. The surgeon cannot afford to be surprised by what is encountered next. The physician, too, needs to know what relationships the organs bear to each other and how this will affect the course of disease. A patient with infection in the appendix, for example, might in some cases show up with pus cells in the urine, because the right ureter passes very near the inflamed appendix. In spite of the pus in the urine, the diagnosis is still appendicitis, not a urinary tract infection. This kind of problem occurs with some frequency, though the organ systems involved are not really functionally related. Thus, the study of regional anatomy is a totally appropriate way for the fledgling doctor to learn anatomy. Regional anatomy is also the *only* way to study on a cadaver. One simply cannot dissect the entire nervous system; then go back and dissect the vascular system; then, the gastrointestinal system; and so on.

Cadaver dissection and regional anatomy are neither necessary nor practical for a course such as this. Instead, we will learn anatomy by *systems* and depend on illustrations instead of cadavers. When studying physiology, it is also necessary to deal with systems, not with regions.

We will correlate the findings in anatomy with the *functioning* of the normal human body, its physiology. Each lecture will concentrate on a particular organ or organ system, for example, the heart. Then, the next lecture will examine the physiology of the normally functioning heart. Finally, to make the connections even more meaningful, we will go into the more common clinical problems that occur when something goes wrong, the pathology of the organ or system. It is these clinical correlations that will make the course meaningful and real. In real life, not everything goes as planned.

For the most part, the lectures are paired, with anatomy first, followed by physiology. In a few cases, this approach is not appropriate, because the physiology of the organ is so much more complex than the anatomy, and to separate the two would be artificial. Lectures Twenty, Twenty-One, and Twenty-Two on the endocrine system are examples.

It would be helpful for the student to own a good dictionary of medical terms such as the one listed in the Suggested Reading. Additionally, having copies of Dr. Netter's *Atlas of Human Anatomy* and Tortora and Grabowski's *Principles of Anatomy and Physiology* at hand would be helpful.

Each lecture ends with questions that should be answered after digesting the material in the lecture. These are, for the most part, complex essay questions and require insight into the anatomy, physiology, and pathology of each system studied. The answers should be easily derived from the lecture notes.

Finally, this course includes a bibliography of suggested reading. The *Atlas of Human Anatomy* by Frank Netter is a classic that has saved the day for thousands of struggling medical students over the years. Its drawings and notes are a must for any serious student of anatomy. *Principles of Anatomy and Physiology* by Tortora and Grabowski should be helpful for its excellent flow diagrams and charts in physiology. Other selections in the list, such as Oliver Sack's *The Man Who Mistook His Wife for a Hat and Other Clinical Tales*, should be fascinating reading for any student who finds, for example, the section on the nervous system to be of exceptional interest.

Although it is certain that this course will *not* prepare you for performing emergency tracheostomy, a wilderness appendectomy, or an informal diagnosis of your neighbor's child's illness, I hope it will excite and inflame an interest in your own body, its processes, and "the ills that flesh is heir to." ■

# Cardiovascular System—Anatomy of the Heart

## Lecture 1

**Now, most courses on the subject begin with a big overview, definitions, relationships of the systems, and we're not going to do that here because I think that can be easily done as we move along through the lectures. Rather we're going to go literally to the heart of the matter and start with the anatomy of the human heart.**

This introductory lecture examines this fist-sized muscle that pumps blood throughout the body. After reviewing the distinction between arteries and veins and discussing the location of the heart, the lecture examines in detail the coverings, layers, and subdivisions of the heart. Finally, it examines the valves that control blood flow into, through, and out of the heart, as well as the conduction system that controls the beating of the heart.

### The heart as a mechanical pump

The heart pumps 1.3 gallons of blood per minute, or 700,000 gallons per year [editor's note: professor correction], at rest and four times that amount at peak exercise. It weighs less than a pound and is the size of a closed fist.

### Types of circulation in the anatomy

- In systemic circulation, arteries carry oxygenated blood from the heart to the rest of the body. Veins then return oxygen-depleted blood to the heart.
- In pulmonary circulation, pulmonary arteries carry oxygen-depleted blood to the lungs, where it is oxygenated and returned to the heart through pulmonary veins.
- Arteries have thick walls and carry blood away from the heart.
- Veins have thin walls and return blood to the heart.

## Location of the heart

Contrary to common belief, the heart is not located on the left side but in the center of the chest in the thoracic cavity, which is bounded by the ribs and the diaphragm. CPR uses the flexible ribs to compress the heart between the sternum and backbone and pump blood in cases of cardiac arrest. Before the advent of CPR, *open cardiac massage* was used.

The thoracic cavity contains three separate enclosures, one for each lung and one for the heart. The *mediastinum* is the heart's enclosure and is the space between the right and left lungs, between the sternum and the vertebral column, and between the clavicles and the diaphragm. The heart's right-to-left axis is between the lungs. Its front-to-back axis is between the sternum and backbone. Its up-and-down axis is between the diaphragm and neck. It sits on the diaphragm.

**In the old days, [pacemakers] had a fixed rate and you had to come in for battery changes, just like you had to come in for an oil and lube job.**

The heart's covering is the *pericardium* ("around the heart"). The *pericardium* is a saccular structure of tough fibrous tissue that holds the heart in place and protects it from overexpansion. The *parietal* ("wall") *pericardium* lines the walls of the heart's enclosure. The *visceral pericardium* is a thin layer that covers and adheres to the heart. The space between the parietal and visceral pericardia is filled with lubricating fluid. Clinical applications: Infections and inflammations of the pericardium (*pericarditis*) can cause friction between the heart and pericardial surfaces, causing severe pain. Trauma can cause blood to fill the pericardium, resulting in pericardial *tamponade* (blood pressing on the heart). Emergency surgery is necessary to relieve pressure on the heart.

## The heart wall

The heart wall has three layers. The outer layer is the *epicardium* (*epi* means "outer"), which is actually the visceral pericardium. It is a shiny, transparent, lubricating layer that is an integral part of the heart wall. The thin inner layer is the *endocardium* (*endo* means "inner"). It covers the muscle and valves, and it continues out into the vessels, where it becomes the *endothelium*.

The middle layer, the *myocardium* (*myo* means "muscle"), is powerful, thick, and tireless. It is a specialized form of skeletal muscle that constantly pumps rhythmically without "instruction" from the nerves or blood. It consists of swirled layers of muscle that envelop the heart so that contraction squeezes the chambers to empty them. There are two sets of muscles: one for the ventricles and one for the atria. The two sets contract independently. The myocardium is an *autonomic* (involuntary) muscle.

**The anatomy of the heart**

The heart is a small, powerful, and untiring organ that pumps blood through more than 50,000 miles of vessels. The nomenclature of "right and left hearts" indicates an artificial separation of heart function. The right (*venous*) heart receives deoxygenated blood from the body and delivers it to the lungs. The left heart receives oxygenated blood from the lungs and delivers it to the body. In the embryo, the right and left hearts start out as two separate tubes that eventually fuse. The right and left hearts are anatomically different and operate at different pressures. The muscles of the left side are four times thicker and more powerful than those on the right side, which pumps blood through a much shorter distance and at lower pressures.

Both sides of the heart have upper and lower chambers. The *atria* (upper chambers) receive blood from the veins and pump it into the ventricles (*atrium* = "waiting room"). The *ventricles* (lower chambers) receive blood from the atria and pump it to the body and lungs (*ventricle* = "little belly"). Because the two circulatory systems are connected, both sides of the heart must pump the same amount of blood. Even the smallest mismatch in circulation causes blood to accumulate in one side of the system, such as in heart failure.

Our four-chambered heart is at the top of the evolutionary line. The right atrium (plural is *atria*) receives deoxygenated blood returning from the body to the heart. Blood returns via the superior and inferior *venae cavae*. The right atrium pumps blood into the right ventricle. The right atrium is a thin-walled, low-pressure system.

The right ventricle receives deoxygenated blood from the right atrium. It pumps this deoxygenated blood to the lungs via the pulmonary artery. The right ventricle is a thin-walled, low-pressure system. The left atrium receives oxygenated blood from the lungs via the pulmonary veins. It pumps blood into the left ventricle. It is a thin-walled, low-pressure system.

The left ventricle receives oxygenated blood from the left atrium. It pumps blood out to the whole body via the aorta. It is a thick-walled, high-pressure system. The intra-atrial septum separates the atria. It contains a closed embryologic opening (the *foramen ovale*). The intraventricular septum separates the ventricles. It contains part of the conduction system (see below).

The valves of the heart prevent backflow. The *mitral* or *bicuspid* ("two leaves") valve is located between the left atrium and left ventricle. It prevents backflow into the atrium during systole. The *tricuspid* valve is located between the right atrium and the right ventricle. It prevents backflow into the atrium during systole. The *chordae tendineae* and papillary muscles hold these valves tightly closed during backflow. The *aortic* ("semilunar," i.e., shaped like a half-moon) valve is located between the left ventricle and the aorta. It prevents backflow into the ventricle during diastole. The *pulmonic* (semilunar) valve is located between the right ventricle and the pulmonary artery. It prevents backflow into the ventricle during diastole.

**Clinical applications**

Damage to the valves can cause *stenosis* (difficulty in opening), *insufficiency* (inadequate closure), or both. *Mitral valve prolapse* (a slight insufficiency) is common in young women. It is often asymptomatic. These conditions can be treated with valve replacements. The coronary arteries supply blood to the heart itself. The left coronary artery comes off the left side of the aorta immediately after leaving the heart. It supplies blood to the muscles of both ventricles, the septum, and some of the left atrium. The right coronary artery comes off the aorta on the right side, level with the left coronary artery. It supplies blood to the right ventricle, left ventricle, right atrium, and septum. Both arteries supply both sides of the heart with plentiful *anastomoses* (places where blood vessels join), which provide redundant blood pathways.

Cessation of coronary artery flow for more than a few minutes causes death of the heart muscle, called *myocardial infarction* ("heart attack" or "coronary"). However, more than 70% of the blood flow must be stopped before tissue damage becomes a problem or any symptoms appear. The coronary veins drain both sides of the myocardium. They empty progressively into the great cardiac vein, the middle cardiac vein, and from there into the coronary sinus in the right atrium.

**Conduction system**

Muscle is inherently conductive, but during embryologic development, about 1% of the heart muscle differentiates to become able to conduct electrical signals almost as well as nerves. Chemical processes in the body produce electrical signals that cause the heart to contract. The *sinoatrial node* is the heart's primary pacemaker. It is located high in the right atrium.

In the old days, pacemakers had a fixed rate and you had to come in for battery changes, just like you had to come in for an oil and lube job, and they had a very short life. Today they're highly sophisticated; they almost never have to be changed.

The *atrioventricular node* is located just above the atrioventricular septum in the base of the right atrium. The conduction system continues into the ventricles via the atrioventricular bundle (the *bundle of His*). The atrioventricular bundle then spreads throughout the ventricles and the septum via the conduction myofibers of Purkinje. This system of conduction is necessary because the myocardium has different inherent contractility in atria and ventricles, and uncoordinated contraction would occur without the conduction system. ■

## Questions to Consider

1. Describe the route of blood flow, starting at the entry into the left atrium through its return to the left atrium.

2. Define the physical borders of the mediastinum.

# Cardiovascular System—Physiology of the Heart

## Lecture 2

**This lecture is going to focus on the physiology of the heart, how it works, how this tremendous pump gets the blood out there, and how it's regulated. And again we'll look at what can go wrong with the mechanisms.**

Now we examine the physiology of the heart, starting with the functioning of the cardiac cycle, in which deoxygenated blood flows into the heart from the body, is pumped out to the lungs for oxygenation, and is then returned to the heart for distribution to the body We also examine the functioning of the heart's conduction system, the functioning of the valves (which produce the heart's distinctive "lub-dub" sound), and possible complications, notably atherosclerosis.

The heart is a powerful pump. It propels blood (a very viscous liquid) through 50,000 miles of vessels. It pumps 1.3 gallons per minute, or 700,000 gallons per year at rest [editor's note: professor correction]. It rests 0.25 second per beat, and it can increase its output by a factor of 5× to 8× under stress. The cardiac reserve is the ratio of maximum output to resting output.

**Systole and diastole**

*Systole* is the active compression, or squeezing, of the ventricles that pushes blood outward to the lungs or the body. *Diastole* is the relaxation of the ventricles when they are filling with blood. The terms *systole* and *diastole* refer only to the ventricles. Atrial systole and diastole exist but are never referred to as simply *systole* or *diastole*.

**Physiology of the cardiac cycle**

Each side of the heart has two chambers. The atria are low-pressure systems, thin walled on both sides. They deliver blood to the ventricles. They contract during diastole to *help* fill the ventricles. Much of ventricular filling is passive; thus, atrial failure decreases cardiac output by only a small percentage (20–30%).

**The ventricles**

The left side of the ventricle is four times thicker and more powerful than the right side. The left (systemic) side is also a high-pressure system. It receives oxygenated blood from the lungs and pumps it out to the body. The right (pulmonary) side is a low-pressure system. It receives deoxygenated blood from the body and pumps it out to the lungs.

**The conduction system**

During embryologic development, 1% of the muscle mass of the heart is designated as *autorhythmic* (self-exciting). These muscles differentiate to form a conduction system. The conduction system establishes the fundamental rhythm. Hormones, chemicals, and nerve impulses can alter the heartbeat strength and heart rate.

The sinoatrial (SA) node initiates the rhythm. It has an inherent rhythm of 60–100 beats per minute. It is located high in the right atrial wall. Impulses spread from the SA node to both atria, causing atrial contraction.

The impulses then spread to the atrioventricular (AV) node, located at the medial base of the right atrium in the atrial septum above the ventricles. AV stimulation sends impulses to the AV bundle of His. The AV bundle of His provides the only electrical connection between the atria and the ventricles. A bundle of His impulses travel through the right and left bundle branches to the conduction myofibers of Purkinje, which conduct impulses to the right and left ventricular muscles, causing ventricular contraction.

**Timing**

The AV node fibers are small, and this delay allows atrial contraction to be completed before the next heartbeat. Conduction speeds up in the AV bundle. The AV node's inherent rhythm is much slower (40–50 beats per minute). Disruption of the SA-AV node sequence will result in a slowed heart rate of less than 60 beats per minute.

Patients with SA-AV node disturbances (bundle branch blocks) can be fitted with pacemakers, which electrically stimulate the heart. Pacemakers can be programmed for the demands of variable pacing. *Valve stenosis* (a tight valve) may cause the valve to overexpand and weaken, increasing the conduction distance. A pacemaker can correct this delay.

**The cardiac cycle**

The first phase of the cardiac cycle consists of the relaxation period. Both the atria and ventricles are in relaxation (diastole). Blood fills the atria from the body and the lungs; when the pressure is high enough, the mitral and tricuspid valves (between the atria and ventricles) open, and blood begins to *passively* fill the ventricles. The atrial contraction assists in filling the ventricles but is not critical. Starling's Law (or Frank-Starling's Law) relates muscular stretching to muscle power. Up to a certain point, stretching a muscle makes it contract with greater power. Atrial contraction stretches the ventricles for greater systolic power.

**The second most common symptom is denial. Almost everybody denies they're having a heart attack. "I'm having indigestion." And many people end up dying with a pack of Rolaids in their hand because they thought they were having indigestion.**

The second phase consists of ventricular filling, which begins. after the valves open. When 30% of the blood is left in the atria, atrial contraction (atrial systole) occurs, emptying the last of the atrial blood into the ventricles. If no atrial contraction occurs, 30% of ventricular filling is lost. An increase in the heart rate can compensate for this loss. The AV valves are still open, and the semilunar valves are still closed.

The third phase consists of ventricular systole (ejection). The ventricles contract, closing the A-V valves. All valves are closed for a fraction of a second. The aortic and pulmonary valves open. Blood is ejected out into the systemic and pulmonary circulations via the aorta and pulmonary artery. When all ventricular force is spent, the aortic and pulmonic valves close, preventing systemic and pulmonary pressure from causing backflow into the ventricles. Once the third phase is complete, the cycle begins again.

**Clinical application**

During ventricular systole, coronary flow is obstructed by squeezing of the myocardium. Therefore, coronary flow to the heart muscle occurs during ventricular diastole, not ventricular systole. During periods of increased heart rate, such as exercise, speed comes at the expense of diastole. In a heart with narrow blood vessels, coronary filling may be compromised to a sufficient extent to cause *angina pectoris* ("pain in the chest"). Patients with only slightly narrowed vessels may be asymptomatic until a significant exertion that sufficiently raises the heart rate. Treadmill, thallium, and echocardiogram tests can identify areas with absent or decreased blood flow by comparing a patient's stressed and unstressed coronary blood flows.

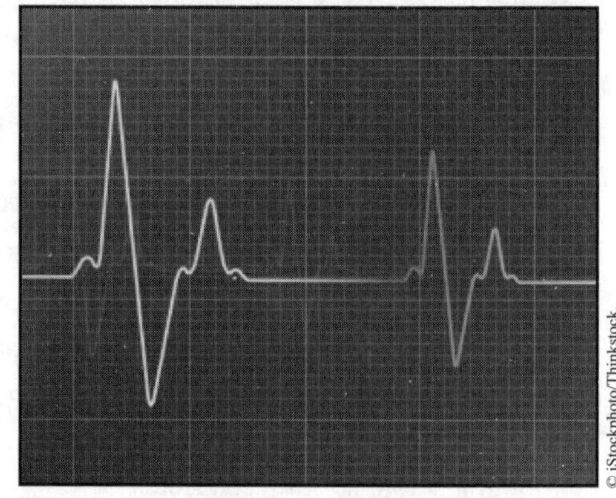



**An electrocardiogram (EKG) provides an electrical picture of the cardiac cycle.**

**Cardiac output**

Stroke volume is measured by the milliliters (ml) of blood pumped out of the left ventricle in one stroke (beat).

**Calculating cardiac output**

Cardiac output (ml/minute) = stroke volume (ml/beat) × heart rate (beats/minute) = ml/minute. At rest, cardiac output (ml/beat) = stroke volume (70 ml/beat) × heart rate (75 beats/min) = 5,250 ml/minute. With moderate exercise, cardiac output = stroke volume (140 ml/beat) × heart rate (150 beats/min) = 21,000 ml/min. Cardiac output can safely go much higher in extreme stress in a trained athlete. A young, healthy person will have a maximum safe heart rate of about 200 beats per minute. The safe maximum heart rate declines with age. A pathological state of arrhythmia can produce rates of up to 300 beats per minute.

**The big picture**

An electrocardiogram (EKG) provides an electrical picture of the cardiac cycle as seen from the surface of the chest. The P-wave is an electrical signature of atrial depolarization. Atrial contraction occurs 0.1 second after the P-wave. The QRS complex marks the beginning of ventricular systole. There is no sign of atrial recovery because it is buried in the electrical portrait of the large QRS wave (atrial recovery is synchronous with ventricular contraction). The T-wave indicates ventricular repolarization, recovery, and rest. The P-R interval is the length of time between the beginning of atrial contraction and the beginning of ventricular contraction. Heart block (A-V dissociation) produces a prolongation or interruption in the P-R interval. Irregularities in an EKG can diagnose many heart problems.

**Symptoms**

The most common symptom of a heart attack is chest pain. The second most common symptom is denial. Almost everybody denies they're having a heart attack. "I'm having indigestion." And many people end up dying with a pack of Rolaids in their hand because they thought they were having indigestion. About 75% narrowing will cause symptoms. And some of these people are correctable merely by diet, some with exercise, and some need more mechanical intervention, the kind of intervention that we've talked about.

**Heart sounds—"lub-dub"**

These sounds are made by blood turbulence as valves close. The sounds are not splashes, because there is no air in the heart, but rather turbulence associated with valve closure. The first sound (lub) comes with the closure of the mitral and tricuspid valves as ventricular systole begins in full. The second sound (dub) comes with the closure of the aortic and pulmonary semilunar valves as systole ends and diastole begins. Mitral and aortic sounds dominate over tricuspid and pulmonic sounds because of the higher pressures in the arterial side (left heart). An important diagnostic tool is the changes of the rhythm and intensities of the sounds in disease. Damaged valves can produce "gallop rhythms," or *murmurs*, which are the sounds of turbulence when valves are damaged. Heart sounds are not heard best directly over the valves but when projected back into the chambers. Stents, angioplasties, and chemicals can repair damaged valves. Bypass surgery reroutes blood around the damaged area. ■

## Questions to Consider

1. Describe the events during the first phase and second phase of the cardiac cycle.

2. Define *cardiac output*. What is *stroke volume*, and how does it affect cardiac output? What is the cardiac output in a man with a stroke volume of 90 ml and a heart rate of 85 beats per minute?

# Cardiovascular System—Anatomy of the Great Vessels

## Lecture 3

**Now we're going to look at the anatomy of the system after the blood leaves the heart. As soon as it's ejected beyond that aortic and pulmonary valve, it's going to enter a network which is comprised of … about 50,000 or 60,000 miles of vessels, an extraordinary amount, and most of those are microscopic.**

This lecture examines the anatomy of the three vessel networks that carry blood from the heart to the body and back again, from the heart to the lungs and back again, and through the portal hepatic circulation of the liver. We begin by identifying and describing the structure of the various vessels that form these networks. Next, we examine separately the major circulatory routes for the blood: arterial and venous systemic circulation, pulmonary circulation, and hepatic portal circulation.

Each vessel belongs to either the right or left side of the vascular system. The left, or arterial, side pumps against the high resistance of the systemic arteries. It distributes oxygenated blood and nutrients to the body. The right, or venous, side is a lower-pressure system that carries deoxygenated blood back to the heart from the lungs.

### Pulmonary circulation

Pulmonary arteries carry deoxygenated blood from the heart to the lungs. Pulmonary veins carry oxygenated blood from the lungs to the heart.

### Overview and definitions

The systemic and pulmonary circulations are both closed-loop systems with the heart as the pump at the center. Systemic circulation carries blood from the heart to arteries, then to capillaries, then to veins, then back to the heart. Pulmonary circulation carries blood from the heart to lung arteries, then to capillaries, then to veins, then back to the heart.

Blood vessels are named for either the specific part of the body they supply or an area surrounding that specific part. Vessels can change names as they run through different parts of the body. The *brachiocephalic* ("arms and head") trunk is the division of the aorta immediately after the aorta leaves the heart. The brachiocephalic trunk splits into the *carotid artery*, which supplies the head, and the *subclavian* ("under the collarbone") *artery*. The subclavian artery becomes the *axillary* ("armpit") *artery*, then the *brachial* ("arm") *artery* as it moves outward along the arm. It splits into the *radial* and *ulnar arteries*, which run along the radius and ulna bones.

On the arterial side, vessels decrease in size in the following sequence: large arteries, medium arteries, small arteries, arterioles, and finally, arterial capillaries (microscopic). Venous vessels *increase* in size as blood returns from venous capillaries to venules, then to small veins, then to medium veins, and finally, to large veins. Veins and arteries generally share the same name; for example, a brachial artery will have a brachial vein next to it. Exceptions include the jugular vein and the vena cava.

Exchange of oxygen, carbon dioxide, nutrients, and waste materials occurs at the capillary level. The capillaries are the only functioning exchange part of the vascular system. The other vessels are merely channels or conduits for the passage of blood.

The smooth muscle of the arteries regulates blood pressure, distribution, and volume. Capillaries comprise most of the 50,000 miles of vessels. The *vasa vasorum* ("vessels of the vessels") feed the walls of the blood vessels themselves.

**Arteries**

The word *artery* is derived erroneously from words meaning "to carry air." Artery walls are composed of elastic tissue and smooth muscle, which is involuntary, out of conscious control. The *lumen* is the hollow center through which blood flows. The walls of arteries are thicker than those of veins.

Conducting arteries are large-sized and contain more elastic tissue than muscle tissue. The largest elastic arteries are the aorta, the carotid, and pulmonary arteries. Elastic recoil maintains pressure between pumping strokes. Back pressure closes the semilunar valves in the aorta. These arteries are larger than 3/4 inch in diameter.

Distributing arteries are small- and medium-sized, and they contain more muscular tissue than elastic tissue. The main function of distributing arteries is to distribute blood flow to parts of the body that need it. They are smaller than 3/4 inch in diameter. They contain smooth muscle in a circular arrangement. These arteries contract when stimulated by the sympathetic nervous system (Lecture 10). These arteries stop bleeding, regulate blood pressure, and shunt blood flow to where it is needed most.

**Arterioles**

Arterioles are very small but visible with the naked eye. They precede the capillaries and contain a good deal of smooth muscle relative to their size. They regulate blood flow to specific capillary beds. They are very numerous (far more than distributing arteries). They are additional important regulators of blood pressure.

Capillaries ("hair-like") are the location of the blood's microcirculation. They are the final smallest pathway for blood. They connect the arterial side with the venous side. The diameter of a capillary's lumen is slightly larger than

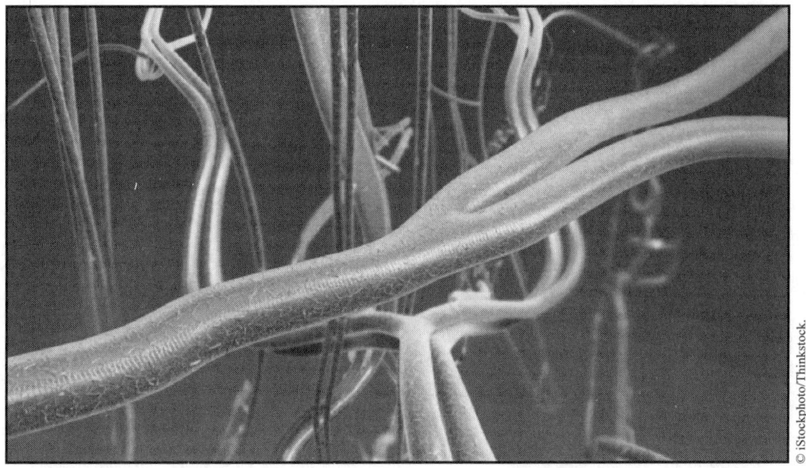



**Blood vessel size varies from the 3/4-inch-wide aorta to the hairlike capillaries.**

one red blood cell (RBC). Every blood cell, therefore, is near the wall of the vessel. This small lumen gives the RBC ready access for gas exchange. The precapillary sphincter is a smooth muscle band preceding every capillary. It opens and closes at different times, shunting blood between various capillary beds.

**The vessels, for all intents and purposes, are conduits. They're just roadways to carry blood, and they have another function for blood distribution and maintenance of blood pressure.**

Every cell in the body is near a capillary. Metabolically active cells (brain, heart, kidneys) have many capillaries. Metabolically inactive cells (bone, tendons, joints) have few capillaries. The lens and cornea of the eye are the only tissues in the body that have no blood supply. They get nutrition by diffusion.

*Anastomosis* refers to the joining of vessels of similar size to supply the same anatomic area. It is a critical safety factor in creating a redundant blood supply for vital organs (heart, brain, intestines). A kidney has only one blood supply vessel, known as an *end vessel*. Each of the two kidneys can function by itself, however, making that system redundant as well.

**Veins**

On the venous side, blood flows in the reverse direction from the arterial side. The venous side returns deoxygenated blood to the heart. Venules receive blood from the capillaries. They empty into progressively larger veins until they reach the heart via the superior and inferior vena cavae. The veins have thinner walls with less elastic tissue than arteries. The inability of lacerated veins to constrict and shut off blood flow can be fatal. Larger veins below the heart have valves to prevent backflow. No valves are necessary above the heart because of gravity. Because the capillaries are under low pressure, they need help to get the blood back to the heart. Muscles massage the veins so that the blood can flow to the heart.

Pooling and clotting in veins can be a problem during prolonged muscular inactivity, such as on a long airline flight. A clot (*thrombus*) can block circulation in that vein. If the clot breaks loose, it will move through the circulatory system. A loose clot is called an *embolus*. Once an embolus moves through the heart, it will encounter smaller vessels and become lodged in a capillary bed somewhere in the body. If the embolus blocks pulmonary capillaries, it is called a *pulmonary embolism*. If the embolus blocks arterial capillaries, it is called an *arterial embolism*. An embolism in the brain is called a *stroke*. Smaller embolisms cause localized cell death; larger embolisms are fatal. The body does not have a sufficient volume of blood for all capillary beds to be open at once. Capillary shunting routes blood to needed areas.

**Clinical applications**

Young, healthy patients can maintain good blood pressure even after losing a large amount of blood. Their ability to shut down capillary beds can cause sudden cardiac arrest as the heart runs out of blood to pump, and a critical line is crossed. This kind of cardiac arrest is almost invariably fatal. Anesthesia can also cause a collapse of blood pressure in young patients, because it relaxes the smooth muscles that close the capillary beds and prevent fatal bleeding and cardiac arrest.

The blood-shunting system is very effective at directing blood to the right areas. The brain's need for blood can shut down every other organ except the heart when blood is at a premium. The venous system contains 60% of the body's blood. Arterial-venous fistulas are places where blood moves directly from arteries to veins, bypassing capillaries. Fistulas can cause heart failure as the heart works harder to supply blood to capillary beds that are not receiving enough.

**Portal circulation**

Portal circulations connect one venous system with another via venous capillaries (rather than arterial-to-venous capillaries). The major portal circulation is through the liver. Another portal circulation exists in the brain around the pituitary gland.

**Hepatic portal circulation**

Venous blood from the digestive system (full of nutrients) circulates to the liver via the following route: The superior mesenteric vein and splenic vein combine to form the portal vein, which lies next to the vena cava. The portal vein enters the liver and divides into the portal capillary bed, a venous-to-venous capillary network. It exits the liver by draining into the inferior vena cava. This portal circulation is an efficient way of delivering blood to the liver for detoxification and nutrient extraction. ■

## Questions to Consider

1. How is the portal circulation different from a systemic capillary bed?

2. How is blood shunted from one organ system to another?

# Cardiovascular System—Physiology of the Great Vessels

## Lecture 4

**We're going to look at the physiology of the "vascular" part of cardiovascular, the blood vessels and the blood itself. The blood is actually an organ. It functions as a unit and has specific jobs to perform, and so we really have to look at the whole blood system as if it were an organ.**

In this lecture, we examine the physiology of the large blood vessels. We will study how these vessels control blood flow, regulate blood pressure, and control bleeding when a vessel is ruptured. We will also examine the composition of blood itself and the functions of each of its constituent parts—plasma, white blood cells, and red blood cells.

### Overview of the anatomy of the blood

Blood is an organ, that is, a system of tissues of different kinds that work together to perform a specific function in the body. The human body contains 5–6 liters (1.5 gallons) of blood, constituting some 8–9% of total body weight. The blood separates into three layers after centrifugation. The densest (bottom) layer is made up of red blood cells, or *erythrocytes* (about 45% by volume). The lightest layer (top) is *plasma* (about 55% by volume). The middle density is white blood cells (*leukocytes*, or the "buffy coat"—about 1% by volume). A hemoglobin reading measures the amount of the hemoglobin molecule in the blood. Hematocrit is the percent by volume of red blood cells.

Plasma (the liquid part) is made of the following:

- Clotting proteins
- Antibodies that fight off infection
- Electrolytes (sodium, potassium, chloride)
- Water

The functions of blood include transportation, protection, and homeostasis.

**Transportation**

- The blood transports oxygen to all of the body's tissues.
- It removes carbon dioxide from tissues.
- Nutrients from the GI tract are absorbed into blood.
- Blood removes wastes from the cells for excretion.
- Blood circulates hormones and other chemicals from sites of manufacture to target organs.

**Protection**

The blood stops bleeding, fights infection, and transports cells and antibodies of the immune system to fight foreign invaders.

**Homeostasis**

- The blood maintains the equilibrium of the *internal milieu*. The body has a very narrow range of acceptable internal conditions.
- It maintains pH (acid-base equilibrium) between 7.1 and 7.4.
- It regulates temperature through shunting mechanisms. Alcohol opens capillary sphincters, especially in the skin, which returns warm blood to the skin in cold conditions. Although this feels pleasant, it draws heat away from the body's core and can result in death.
- It regulates total fluid and electrolyte balance.

**Blood composition**

Ninety-two percent of total body weight is water, bone, and other tissue. The tissue is also mostly water. Whole blood makes up about 8% of total body weight. Plasma makes up 55% of blood, and it is 95% water. About 45% of blood comprises formed elements, which are the erythrocytes, the leukocytes, and the platelets. Erythrocytes are the vast majority.

**Erythrocytes (red blood cells)**

Erythrocytes capture oxygen in the hemoglobin molecule, release it to cells, and capture carbon dioxide for disposal. Erythrocytes are biconcave disks, which maximize gas-exchange ability while minimizing size. Only immature erythrocytes have nuclei; patients with anemia will have too many erythrocytes with nuclei. Each red blood cell contains about 300 million hemoglobin molecules, which are iron-based and have a very high affinity for oxygen. Oxygen release and carbon dioxide capture are pH-based. Hemoglobin's affinity for carbon monoxide (CO) is 200 times greater than for oxygen. CO can replace oxygen in the blood and cause death. CO poisoning can be treated by having the patient breathe 100% oxygen to replace the CO.

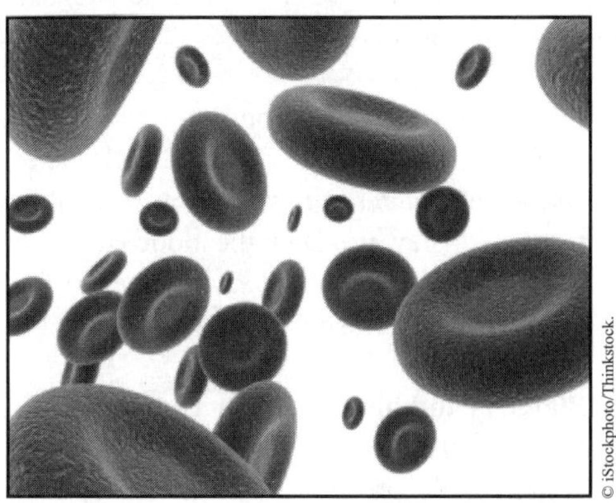



**Erythrocytes, or red blood cells, carry oxygen.**

**Leukocytes (white blood cells)**

There are many different kinds of leukocytes with different functions. The hematopoietic stem cell is the origin of all leukocytes. A stem cell is a cell that can turn into any type of cell when directed by chemical signals during embryonic development. There are several kinds of white blood cells.

- *Eosinophils* are involved in allergy responses.

- *Neutrophils* are the main bacteria-fighting elements and have different possible nucleus shapes.

- *Basophils* are involved in allergy processes.

- *Monocytes* are "wandering macrophages" that travel the body eating dead cells and waste.

- *Lymphocytes* play an important role in the immune system.

- *Megakaryoblasts* become megakaryocytes, which break up into platelets.

Leukocytes alert the body to "stranger danger" and release lysins and perforins, which destroy foreign cells. They can move out of blood vessels and attack danger in tissues as well. Leukocytes that leave the blood cannot reenter it. They also keep the blood sterile. Bacteria in the blood (*bacterimia*) is very serious, as is *viremia* (viruses in the blood). Most life-threatening is *septicemia* (organisms multiplying in the blood).

**Platelets and hemostasis mechanisms**

Platelets are the first line of defense in the blood's complicated clotting system. In the process known as *chemotaxis*, chemical signals attract platelets to a trauma site. They aggregate and form a barrier to blood flow.

**Platelet plug formation**

- Platelets become sticky and adhere to the torn vessel at the site of bleeding.

- The platelets release chemicals; there is a further spasm of smooth muscle and more narrowing of the hole.

- Sticky platelets call for more platelets until a plug is formed.

- This is only the first step; the platelet is the paramedic of the bloodstream.

A clotting cascade then begins forming a more permanent clot. This is a complicated cascade, with a very delicate balance, that produces jelly-like clot forms that will later be dissolved by the body. The clotting cascade is complicated to prevent blood from clotting inside healthy blood vessels. Disseminated intravascular clotting is a disease in which blood clots inside healthy vessels. Because all the clotting factor is used up, patients will, paradoxically, have widespread bleeding. Hemophilia is a disease in which the patient lacks a clotting factor. Wounds will clot initially, but the patient can bleed to death later without treatment. *Thrombi* (clots) and *emboli* (clots moving in the blood) are unwanted clots that can cause death.

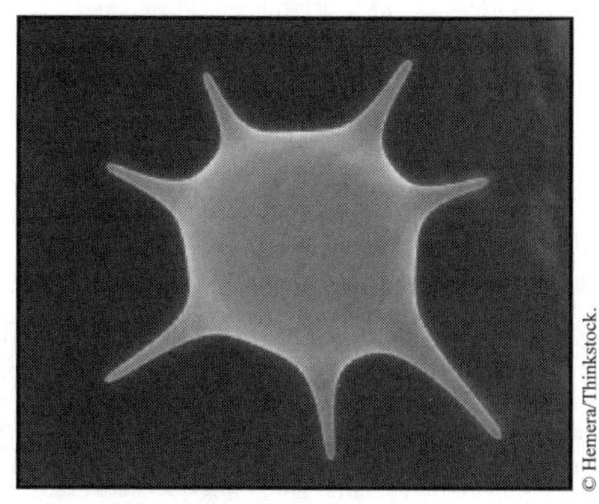


**Platelets initiate blood clotting.**

## Control of blood pressure

The pacemaker system in the heart controls only the *rate* of blood flow. Cardiac output is varied by rate and stroke volume. Baroreceptors in the carotid artery and aortic arch measure blood pressure. Baroreceptors report drops in blood pressure to the brain, which will take any measures necessary to increase the pressure.

The need for brain blood flow will override almost any other signal. The brain does not care *why* pressure is low. It will act immediately to increase rate, stroke volume, and so forth. Long-term solutions come later. Causes (dehydration, bleeding, drugs) are later problems. The brain can be fooled. Pressure on baroreceptors of the neck signals falsely high pressure. The brain responds to lower it, and fainting may follow. Paroxysmal atrial tachycardia causes the atrium to beat at 300 beats per minute. Pressure on the baroreceptors will cause the brain to send a signal to the heart to slow down.

## Shock

Shock is a state in which the body cannot maintain sufficient cardiac output to supply oxygen and nutrients to the cells and remove wastes. *Shock* is not synonymous with either "stroke" or "emotional shock." Shock is defined as a sustained systolic pressure of less than 80 mm of mercury. It causes the blood pressure and flow for the affected person to become too low. If shock persists too long, it becomes irreversible, and death will follow.

**Types of shock**

- *Hypovolemic shock* involves a decrease in blood volume. This decrease can result from blood loss (e.g., as a result of trauma or GI bleeding) or from dehydration (e.g., as a result of diarrhea, vomiting, lower fluid intake, or excessive sweating).

- *Endotoxic shock* results from infection with certain organisms, especially gram negative bacteria.

- *Cardiogenic shock* results from malfunction of the heart (either a nonfatal heart attack or arrhythmia).

- *Neurogenic shock* is a result of the brain trying to shut itself down, for example as a result of a sudden, unwanted visual image.

**Stages of shock**

Stage I is compensated non-progressive shock in which there is no cell death or permanent damage. This stage, which occurs after mild blood loss, is typified by transient fainting. Body responses are totally adequate.

Stage II is decompensated progressive shock in which cells are injured. It results from more severe blood loss (about 25% of blood volume) or severe dehydration (e.g., as a result of cholera). It cannot be reversed without outside help.

Stage III is irreversible shock. The organs are damaged, the blood becomes acidic, and death usually occurs within hours. No treatment is possible. ■

## Questions to Consider

1. Define *shock*.

2. What are the three stages of shock, and which (if any) are treatable?

# Respiratory System—Anatomy of the Lungs

## Lecture 5

**Today, our lectures are going to take us into a new system, the respiratory system. And we put it here because it has a very natural fit both anatomically and physiologically with the cardiovascular system.**

The respiratory system brings oxygen to move carbon dioxide from the blood so that the heart can pump it out to the rest of the body and to all the cells. This lecture examines the anatomy of the respiratory system, especially of the lungs. After studying the integration of the respiratory system with the circulatory system, we review the anatomy of the structures through which air enters the body and passes into the lungs. Next, we examine in detail the anatomy of the lungs themselves—their sections and covering and the structures in which exchange of gases with the blood takes place.

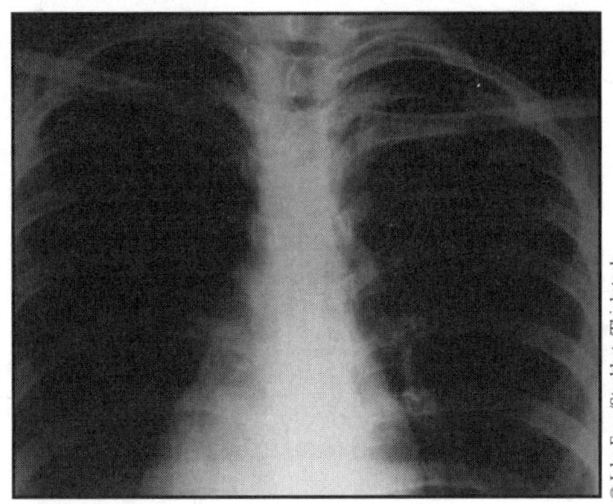



**The respiratory system under X-ray.**

The respiratory system performs the following functions:

- It takes up oxygen for transport in the blood to the cells.
- It expels carbon dioxide waste from cellular metabolism.
- It helps regulate acid-base balance through the bicarbonate buffer system.

**Anatomy of the respiratory conducting system**

- The respiratory conducting system brings air to the respiratory portion.
- The respiratory and digestive systems are connected and share the conducting system of the mouth and throat.
- The mouth is a conducting system to the larynx. It moisturizes air.
- The nose is a conducting system to the larynx. It moisturizes and warms air and removes particulate debris.
- The mouth and nose work together to provide backup volume and redundant sources of air supply.
- The nose is lined with mucous membranes that have folds called *turbinates* or *conchae*. Turbinates provide extra surface area to warm and moisturize air.
- The mucous covering the turbinates is sticky and catches dirt and other foreign elements, which are then swallowed or blown out.
- The *nasopharynx* (*pharynx* means "gulf" or "throat") has the same functions as the nose, but it has a larger volume.
- The *oropharynx* is the back of the mouth.
- The *epiglottis* covers the trachea during swallowing.
- The *hypopharynx* is the lowest part of the pharynx.

**The larynx**

- The *larynx* is the "voice box"; it functions only to produce sound.

- The lips and tongue form the sounds into words, songs, and screams, but they do not produce sound as such.

- The larynx is at the midline of the neck, halfway between the clavicle and jaw.

- The *thyroid cartilage* is known as the Adam's apple.

***Vocal folds* (also known as "vocal cords")**

- These structures tighten and move toward the midline to produce sound by vibrating when air passes by them.

- Tighter tension results in higher pitch.

- The folds are tensed by the intrinsic muscles in the voice box.

- Paralyzed cords meet in the midline, during acute phase.

- Male hormones make these structures thicker and the voice deeper.

- The cricothyroid muscles move the larynx forward and down, raising the pitch of the voice.

- The vocal cords and cricothyroid muscles are controlled by the laryngeal nerves. Paralysis of one nerve leads to hoarse speech; paralysis of both nerves can lead to complete airway obstruction and death.

Emergency tracheotomy surgery to bypass an obstructed airway is commonly regarded as easy or uncomplicated. On the contrary, it presents many problems even for trained surgeons. An elective tracheostomy is the semipermanent or permanent addition of an artificial airway to assist breathing. (The suffix *–ostomy* implies some permanence, while *–otomy* implies cutting open, then closing again.)

**Pleura and pleural spaces**

The pleura are sacs made of strong connective tissue that line the chest wall (the parietal pleura) and the lungs (the visceral pleura). The inferior pulmonary ligament joins the pleura. The space between the parietal and visceral pleura is the pleural cavity. It is filled with a lubricating fluid.

**Anatomy of the conduction portion**

The trachea is a tubular structure just beneath the larynx. It is made of stacked rings of cartilage with fibrous spaces. It divides in the upper chest into the right and left main stem bronchi. The main stem bronchi carry gases to the right and left lungs. They divide into smaller and smaller segmental bronchi. The whole is called the *tracheobronchial* tree. The bronchi divide finally into *bronchioles* (small bronchi).

The lungs are divided into lobes (upper, middle, and lower). These lobes are divided by fissures into segments. The segments are served by the smaller divisions of the bronchi. The lung is not a hollow balloon-like structure but a spongy mass of tissue consisting of millions of microscopic balloons called *alveoli*, surrounded by capillaries.

**Clinical application**

Removing a lung is much easier than removing a specific bronchial section. However, smokers and other patients who need a lung removed often do not have enough reserve lung capacity to survive this type of surgery.

**Anatomy of the respiratory portion**

Bronchioles end the conducting system, and alveolar ducts, alveolar sacs, and alveoli begin the respiratory exchange portion.

The alveoli are the home of pulmonary (external) respiration.

- Exchange of gases occurs in alveoli.
- Alveolar ducts direct air from the bronchioles into the alveolar sacs.
- Alveolar sacs are clumps of alveoli with a common opening.

- Each alveolus is only one cell thick.

- Pulmonary capillaries are directly next to alveolar cells, and they also are only one cell thick.

- Carbon dioxide passes from the pulmonary capillary bed into the alveolus lumen for exhalation.

- Oxygen passes from the inspired alveolar air into the pulmonary capillary bed and into the blood.

- Emphysema breaks down the walls between sacs and reduces the amount of gas that can be exchanged.

Pulmonary ventilation is simply the movement of air. Pulmonary respiration is both the movement of air and the exchange of gases with the blood.

**Clinical applications**

Diseases such as asthma can cause the bronchioles to close, trapping air inside the lungs and preventing respiration. Also, damage to the tracheal and bronchial cilia can prevent foreign matter from being ejected, causing disease. ■

## Questions to Consider

1. Define the *pleural space* anatomically.

2. Would tracheostomy help in the case of a severe asthma attack? If so, why? If not, why not?

# Respiratory System—Physiology of the Lungs

## Lecture 6

**We're back today to talk about the physiology of the respiratory system and see what really happens now once the air is into the lungs, how it gets there, and how chemically we're going to get our gases exchanged.**

In our continuing study of the respiratory system, we examine the four areas in which respiration occurs and the consequences of blocking this respiration (i.e., suffocation). Next, we review the physiology of the movement of gases into the lungs and of their exchange with waste gases in the blood at the level of the alveoli. Finally, we turn to the central respiratory centers that control the pace of respiration, and we examine some common respiratory disorders and their treatment.

The larynx, the trachea, the bronchi, and their smaller divisions perform pulmonary ventilation, which is simply the movement of air. The bronchioles and alveoli perform pulmonary respiration, which is the exchange of gases.

**Four areas of respiration**

- In pulmonary ventilation, room (or outside) air is taken into the body. Room air is about 80% nitrogen and 20% oxygen, with trace gases.
- External pulmonary respiration involves the exchange of gases within the lungs, in the blood of the capillaries of the alveoli.
- Internal (tissue) respiration involves the exchange of gases between cells and blood in the capillaries.
- Cellular respiration involves metabolic processes within the cell.

Inspiration is taking air into the lungs; expiration is pushing air out of the lungs. The term *expiration* also means "to die."

**Movement of gases**

When the mouth and nose are open, the air pressure in the lungs equals the atmospheric pressure. The pressure of the space between the visceral and parietal pleura is slightly less than the atmospheric pressure. Inspiration contracts the diaphragm and increases the diameter of the thoracic cavity, decreasing the pressure between the lung and the parietal pleura. The lowered pressure of the pleural space draws air into the lungs. On expiration, relaxation of the diaphragm increases the pressure in the pleural space and expels air from the lungs. Expansion and contraction of the flexible ribs and chest aid inspiration and expiration.

The accessory muscles of respiration aid in times of pulmonary stress, such as during exercise.

- The external intercostal muscles connect the ribs and the collarbone, pulling the ribs up during respiration.

- The sternocleidomastoid muscle helps raise the first rib, sternum, and collarbone by pulling against the head.

- The scalene muscles also pull the ribs up against the collarbone.

- The rectus abdominus and internal intercostal muscles pull the ribs down, aiding expiration.

**Alveolar capillary air exchange**

This exchange is an example of pulmonary (external) respiration. It involves the exchange of gases between the air in the alveoli and blood in the pulmonary capillaries. The movement of this exchange is dictated by gas laws, which say that gases move from areas of high pressure to areas of low pressure. Because the partial pressure of oxygen is higher in the lungs than in the alveoli or capillaries, oxygen pushes into the capillaries and is captured by hemoglobin in the blood. Because the partial pressure of carbon dioxide is higher in the capillaries than in the lungs, carbon dioxide pushes through the alveoli into the lungs. Carbon dioxide also comes from buffer systems in plasma (not only from red cells) that rid the body of excess acid.

Suffocation could, therefore, occur at any of the following levels:

- Pulmonary suffocation results from closure of the nose and mouth, or throat, by a foreign body. The consequence is strangulation.
- Alveolar suffocation results from water in the alveoli of the lungs.
- Tissue suffocation results from obstruction of flow in the capillary bed.
- Cellular suffocation results from metabolic poisons, such as cyanide.

**Common clinical obstructions**

There are very few common clinical obstructions of the upper airway (the nose and mouth) because of the existence of dual passageways and the ease of clearing them.

In the larynx, the epiglottis prevents food from entering the airway on swallowing. It can malfunction if one breathes while swallowing (e.g., when talking). *Epiglottitis* (viral infection of the epiglottis) can very rapidly cause death if the airway is closed by swelling. The involuntary gag reflex protects this pathway. Violent spasms of coughing serve to expel foreign matter. The lungs treat the presence of foreign matter as a serious threat.

*Laryngospasm* is an over-response that causes the vocal cords to close. Alcohol, anesthesia, and other drugs can decrease or eliminate the gag and cough reflexes. The Heimlich maneuver is used to expel impacted foreign matter.

**Tracheo-bronchial tree**

- Cilia in the lining cells of the upper airway beat toward the mouth to remove small foreign material.

- Loss of cilia does double harm to smokers, because secretions cannot be cleared and carcinogens stay in contact with the lining cells longer.

- Lung cancer has an extremely high fatality rate and is easily preventable.

- A foreign body impacted here may be rapidly fatal; it is hard to gain emergency access. Examples include the intake of water during drowning or the aspiration of vomit.

**Loss of cilia does double harm to smokers, because carcinogens cannot be cleared from the lungs' lining.**

**Central control of respiration**

This central control is both conscious and unconscious (voluntary and involuntary). Breath can be controlled to a great extent but not to the point of death. There are several centers of respiratory control, all of which are driven by small changes in blood chemistry.

- The medullary rhythmicity center establishes the basic normal rhythm (12–15 breaths per minute at rest). It has separate inspiratory and expiratory centers.

- The apneustic area controls breath holding.

- Pneumotaxic areas regulate the brain's coordination of respiration.

**Pneumothorax (air in the chest)**

Air entering the pleural space between the lung and the chest wall causes loss of the vacuum. The air comes from outside of the chest (e.g., through a hole in the chest wall) or from the inside (e.g., a hole in the lung). The lung collapses (partially or completely) and gas exchange does not take place on that side. Treatment involves the insertion of a chest tube and suction. Treating a hole in the chest wall is much easier than treating a hole in the lung. Causes of pneumothorax include trauma (guns, knives, automobiles), spontaneous pneumothorax, a ruptured bleb (caused by emphysema), or hyperinflation of the lung during anesthesia.

**Pulmonary edema (fluid in the lungs)**

Cardiogenic edema is caused by failure of the left heart, which causes blood to back up into the lungs. Right heart failure causes edema in the body. Fluid in the lungs fills the alveoli and prevents gas exchange. ■

## Questions to Consider

1. What muscles are involved in respiration during severe exertion? How is that different from respiration at rest?

2. What is a *pneumothorax*? Describe the events that occur in the chest after a gunshot wound that creates a hole in the chest wall.

# Nervous System—Anatomy of the Brain

## Lecture 7

**Today we're going to talk about the brain. We're going to begin what is one of the largest series of lectures in this whole course devoted to the study of the nervous system. Now, the brain is where most of us believe that it all happens; it's one of the most complex organs in the body, probably one of the least fully understood.**

In this, the first of seven lectures on the nervous system, we start by examining the brain: its principal components (the outer coverings, the cerebrospinal fluid, and the blood vessels) and its main anatomical divisions (the brain stem, the cerebellum, the diencephalon, and the cerebrum). Next, we examine the divisions of the cerebrum (two hemispheres, each with four lobes) and the functional areas of the cerebral cortex.

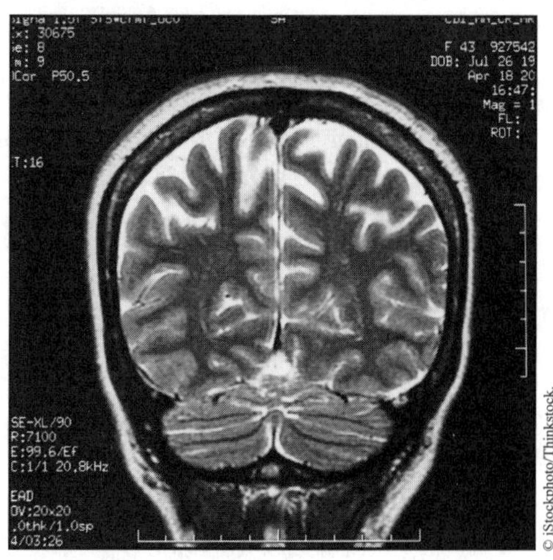


**The largest part of the human brain is the cerebrum, which divided into two hemispheres (left and right) and four lobes.**



The brain is among the largest organs in the body, weighing about 3 pounds and containing more than 100 billion cells. It constitutes just 2% of the body's weight but consumes 20% of the body's blood supply. Virtually all of the activities of the rest of the body are devoted to the care and protection of the brain. The brain is covered and protected by a rigid, bony case—the skull. The skull is composed of two layers of bone with air spaces between them. This makes the skull lighter and more protective than solid bone.

**Principal components of the brain**

- The *meninges* are the coverings of the brain.

- The *dura mater* ("hard mother") is the outer and toughest covering. It also separates most of the two halves of the brain.

- The *arachnoid* ("spiderlike") is the middle layer. It circulates cerebrospinal fluid and blood.

- The *pia mater* ("delicate mother") is the innermost covering. It adheres to the brain.

- The *arachnoid villae* secrete cerebrospinal fluid (CSF) in the brain's *sinus* (open spaces). CSF covers the brain and spinal cord.

- CSF is produced by the choroid plexuses, located in the walls of the ventricles. It is reabsorbed through the arachnoid villae in the dural venous sinuses. It has the following functions:

    - It provides mechanical protection for the brain and spinal cord, and it provides buoyancy.

    - It provides chemical protection: It is the optimal environment for neuron function.

    - It circulates nutrients to parts of the brain, and it is an exchange area for wastes.

The brain requires a slightly different operating environment than the body. The blood-brain barrier separates the two systems.

**Clinical application**

Skull fractures almost always tear the dura mater, resulting in one or more of three typical clinical situations:

- A *subdural hematoma* is a pool of blood under the dura that puts pressure on the brain. Subdural hematoma is common in the elderly, because the brain shrinks with age and a large hematoma can develop without symptoms.

- An *epidural hematoma* is a pool of blood between the dura and skull.

- A *subarachnoid hematoma* is functionally an intracerebral hemorrhage and is usually untreatable once it occurs, because draining it would cause more brain trauma.

**The ventricles**

The ventricles of the brain are reservoirs of CSF. Two lateral ventricles are located in the cerebral hemispheres. The third ventricle is in the midline between the halves of the thalamus. The fourth ventricle is between the cerebellum and brain stem. Blockage of CSF circulation can lead to hydrocephalus, which can be corrected with mechanical shunts and valves.

**Circulation in the brain**

Blood is supplied to the brain through the internal carotid artery and the basilar artery (formed by the junction of the right and left vertebral arteries). The internal carotid and basilar arteries join to form the *circle of Willis*, from which branch the anterior cerebral artery, the middle cerebral artery, and the posterior cerebral artery. The circle of Willis provides anastomoses for the two sides of the brain and backup for failure of one or more arteries. Vertebral arteries join in the head to form the basilar artery. An incomplete circle of Willis exposes parts of the brain to possible damage in the case of occlusion of a vessel.

Interruption of blood flow to the brain for seconds causes loss of consciousness. An interruption for 1–2 minutes can cause brain damage. An interruption of more than 4 minutes can cause death.

Glucose stores in the brain are limited; thus a drop in blood glucose can cause confusion, convulsions, loss of consciousness, and death. The diving reflex shuts off circulation to unneeded parts of the body on immersion in very cold water. This can lead to survival for periods much longer than 4 minutes without brain damage.

**Anatomical divisions of the brain**

The brain is almost fully developed at birth, lacking mostly insulation. The brain stem connects the spinal cord to the *diencephalon*. The *medulla oblongata* starts at the *foramen magnum* ("the big hole").

- It controls cranial nerves VIII, IX, X, XI, and XII.
- It contains the cardiovascular center (heart rate) and the respiratory center (sneezing, coughing, hiccoughing).
- Motor and sensory nerves cross sides in the medulla oblongata.
- Within it are all ascending (sensory) and descending (motor) tracts.

The pons ("bridge") controls the nuclei for cranial nerves V, VI, VII, and VIII. It is also the location of several important structures:

- The motor and sensory tract convergence.
- Fibers connecting the right and left sides of the cerebellum.
- The pneumotaxic and apneustic centers.

The *cerebellum* coordinates input from the sensory organs of proprioception and movement and ensures that intention and actual movement are the same. It makes possible coordinated complex motor function, such as dancing, playing the piano, and engaging in sports. Recent research indicates that the cerebellum may perform many more functions than previously thought. Certain unlearned reflexes may have come from ancient conditions and may have been "hardwired" into the brain during evolution.

The *diencephalon* is found between the midbrain and the cerebrum. The thalamus is the major center for receiving input to the brain from the periphery and the brain stem. It is the center for acquisition of knowledge, awareness, emotion, and memory.

The *pineal gland* secretes melatonin, a sleep inducer, and sets the internal biologic clock. The *hypothalamus* is located under the thalamus. It relays reflexes related to smell. It secretes hormones for regulating the hormones of the anterior pituitary gland. It is the major regulator of homeostasis; it receives input that is interpreted to correct for changes in osmotic pressure, hormone concentrations, and blood temperature. It controls the autonomic nervous system (heart rate, respiration, gastrointestinal tract, urinary bladder) and body temperature. It not only affects emotions of aggression, pain, and pleasure; it regulates thirst and hunger responses and controls and it regulates rhythms of sleep and wakefulness.

The cerebrum contains the bulk of the brain's mass. The cerebral cortex is made of an outer layer of gray matter composed of more than 1 billion neurons. Beneath the gray matter is white matter, made up of myelinated and unmyelinated axons connecting neurons from different parts of the brain. The two hemispheres are separated by an extension of the dura mater called the *falx* ("sickle-shaped") *cerebri*, and they are connected by the *corpus callosum* ("hard body"). Each hemisphere is divided into four lobes: frontal, parietal, occipital, and temporal. ■

## Questions to Consider

1. What are the three layers of tissue that cover the brain? Describe the function of any two of them.

2. How does the circle of Willis provide insurance against brain infarction?

# Nervous System—Physiology of the Brain

## Lecture 8

**Today we're going to continue our journey through the nervous system, and we're going to begin to focus on the components at a microscopic level, as well as the physiology, not so much where things are and what they look like, but really how they work and how all of this system gets integrated.**

The nervous system and endocrine system function to maintain homeostasis. After considering how the nervous and endocrine systems work together for this purpose, we'll review the functions of the nervous system and the varieties and functions of nervous tissue, as well as the main divisions and subdivisions of the central and peripheral nervous systems. We will distinguish afferent from efferent nerves, and we will describe the functional categories of cranial and spinal nerves.

The nervous system sends electrical signals to individual cells. It does so very rapidly (in milliseconds). Duration of action is very short. The nervous system targets specific organs for action. Specific receptor organs at the end of each neuron receive these signals. The endocrine system secretes hormones. It functions much more slowly than the nervous system (seconds to hours). It is less specific. It reaches all of the body's cells via the bloodstream. Duration of action is longer than the electrical signals.

The nervous and endocrine systems provide redundancy and different types of action. Some actions, such as adrenaline production, can be performed by either or both systems.

**Functions of the nervous system**

- Nervous impulses travel in two directions, and each direction has a separate set of nerves.
- *Efferent*, or descending, nerves send impulses away from the brain. These are also called *motor nerves*.

**[Dr. Wilder Penfield] could anesthetize the scalp ... open that up, and then open up the brain case, the skull ... and expose the brain to cut through that dura mater, the hard covering, and look right at the brain.**

- Sensory impulses that ascend the nerves to the brain are called *afferent impulses* and move along *afferent nerves*.
- Efferent impulses produce action in distant organs.
- Afferent impulses deliver information from peripheral nerves to the brain.
- The integrative function analyzes information input from afferent impulses, files these data, and stimulates action in response to sensory input.

**Kinds of nervous tissue and their functions**

*Neurons* ("nerve cells") initiate and relay electrical nervous impulses toward and away from the brain. Neurons have three basic parts:

- The cell body
- *Dendrites* ("trees"), which receive incoming messages
- *Axons*, which deliver outgoing messages

Nerves can have any arrangement of dendrites and axons, including one input to many outputs (*divergence*), many inputs to one output (*convergence*), and one input to one output.

Some axons are *myelinated* (insulated) for rapid impulse delivery. These axons display *saltatory* (jumping) conduction. Other axons are unmyelinated. Myelin sheaths develop as a baby grows; disease can interrupt this process.

*Neuroglia* ("nerve glue") are connective nerves that hold the structure of the nervous system together. These cells nurture the neurons. They are much more numerous than the neurons and are regenerated over time. Because they can regenerate, their stem cells, not the neurons, are the source of most cancerous cells in brain cancer.

**Divisions of the nervous system**

The main divisions of the central nervous system (CNS) are the brain and spinal cord. The central nervous system contains the longest cells in the body. The main divisions of the peripheral nervous system (PNS) are the cranial nerves and the spinal nerves. The spinal nerves connect the neurons of the CNS to the effector organs (muscles and glands) or the sensory organs (pain, heat, proprioception).

The PNS is divided further into:

- The somatic nervous system (SNS), which is voluntary and conducts impulses to the skeletal muscles.

- The autonomic nervous system (ANS), which is involuntary and conducts impulses to the internal organs. The ANS is divided into:

    - The sympathetic nervous system, which controls increased survival activity through the "fright-fight-flight" reactions.

    - The parasympathetic nervous system, which controls rest, recovery, and relaxation.

**Other nerve classifications**

- Afferent (*ferre* means "to carry"), which carry impulses toward the central nervous system (sensory).

- Efferent (meaning "away from"), which carry impulses away from the brain toward the periphery (motor).

**Functional categories of the spinal and cranial nerves**

- The general somatic sensory neurons (afferent) transmit impulses of pain, temperature, touch, proprioception, and vibration. These impulses are transmitted to the brain from the skin, joints, muscles, and cranial nerves.

- The special somatic sensory neurons (afferent) transmit impulses for vision, hearing, and balance. These impulses are transmitted from the organs of special sense toward the brain via the cranial nerves.

- The general visceral sensory neurons (afferent) transmit information impulses from the autonomic nervous system about the condition of the viscera. The impulses are transmitted to the brain along the cranial nerves and the spinal nerves.

- The special visceral sensory neurons (afferent) transmit impulses of taste and smell from the organs of special sense to the brain along the cranial nerves.

- The general somatic motor neurons (efferent) transmit outgoing impulses to the skeletal muscles from the brain along the cranial nerves and the spinal cord.

- The general visceral motor neurons (efferent) transmit impulses from the CNS to the viscera in the autonomic nervous system (smooth muscle, cardiac muscle, and glands) along the cranial and spinal nerves.

- The special visceral motor neurons (efferent) transmit impulses from the CNS along the cranial nerves to the skeletal muscles of facial expression, larynx, and pharynx.

**Summary of the functions of the brain's divisions**

The brain's surface is convoluted, with *gyri* (the singular is *gyrus*) and *sulci* (singular, *sulcus*; "shallow grooves") adding surface area for neurons. Brain mapping has identified gyri that control specific functions, such as vision, touch, and speech.

Certain types of epilepsy are caused by scarring of brain tissue, which can disrupt the normal function of, for example, a motor-control section.

Misfiring neurons can cause other neurons to fire, triggering a grand mal seizure. Petit mal seizures and narcolepsy are other results of scarring. Temporal lobe seizures (psychomotor epilepsy) are caused by signals from the temporal lobe and can be violent.

The brain tissue has no pain receptors. Surgeons can operate on a brain while the patient is awake without anesthetizing the brain itself. Direct electrical stimulation of the brain while the patient is conscious allows brain functions to be mapped. The *homunculus* ("little man") is a visual map of bodily functions superimposed on the structure of the brain itself. It indicates the relative importance of various functions by their size. Positron emission tomography (PET) scans show cell function in the brain and can indicate which areas are used for more complex activities, such as writing, playing music, and engaging in mathematical analysis. ■

## Questions to Consider

1. What is the difference between the sympathetic nervous system and the parasympathetic nervous system?

2. How are the endocrine system and the nervous system different in their regulation of homeostasis?

# Nervous System—Spinal Cord and Spinal Nerves

## Lecture 9

**Today we're going to talk about the major pathways that get impulses to and from the brain to the rest of the body. And we're going to be dealing with the spinal chord and the spinal nerves.**

This lecture examines the anatomy and functions of the spinal cord, including the outer coverings and layers of the cord, the spinal fluid, and the internal components of the cord. Next, we review the reflex arc, which allows the body to react rapidly to changes in the external environment, bypassing the brain. Finally, the lecture examines the categories and locations of the spinal nerves.

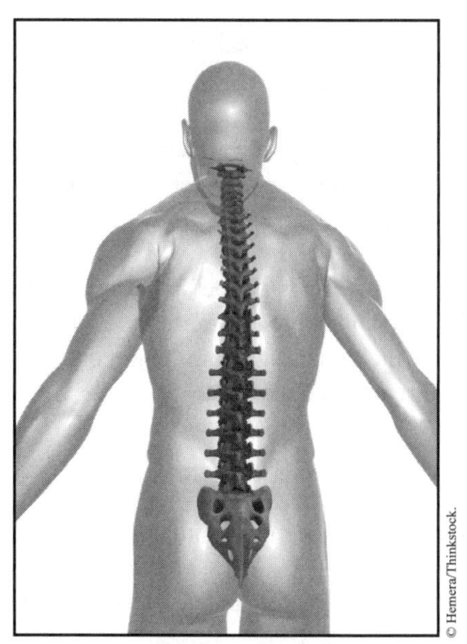

© Hemera/Thinkstock.

**The spinal cord connects the brain with the rest of the body.**

The spinal cord is still part of the central nervous system. It begins at the foramen magnum, the large hole at the base of the skull. It connects the brain with the rest of the body. Ascending (afferent) fibers bring sensory input to the brain from the body. Descending (efferent) fibers send motor signals to the body. The spinal reflex arc receives sensory signals from the body and relays immediate responses via the motor neurons to take action without intervention from the brain. The brain modulates the reflex, which is why a severed cord (interrupting the brain's modulation) causes hyperreflexia.

**Spinal cord anatomy**

The spine runs from the base of the skull to the tailbone. At the top of the spine, the atlas holds the skull up and the axis allows it to rotate. It is divided into four sections.

- Eight spinal nerves exit the *cervical* (neck) section.
- Twelve spinal nerves exit the *thoracic* (chest) section.
- Five spinal nerves exit the *lumbar* (back) section.
- Five spinal nerves exit the *sacrum*, in which the vertebrae are fused together.

The spine is curved in an anterior-posterior (front to back) direction, and the curves alternate between convex (*kyphosis*) and concave (*lordosis*), allowing the spine to act as a spring.

The coverings of the spinal cord are completely analogous to those of the skull and brain. The vertebral column is analogous to the skull. The meninges of the spinal cord are analogous to the meninges of the brain. There are three meninges:

- The dura mater
- The arachnoid
- The pia mater, which is adherent to the cord

The spaces are the same (epidural, subdural, and subarachnoid). The CSF in these spaces is continuous with the CSF of the brain. Thus CSF analysis from a spinal tap gives information about the brain as well as the cord. Anesthesia can be introduced into these spaces to aid surgery. However, anesthesia that ascends too high in the spine can stop respiration. Missing CSF can cause the brain and cord to sag and produce severe headaches. Spinal nerves near the end of the spine simply fan out at the end of the cord, forming the *cauda equina* ("horse's tail").

### The spinal cord

The spinal cord is about 1 inch in diameter. Gray matter (butterfly shaped) is in the center. It is divided into posterior (back, or dorsal), anterior (front, or ventral), and lateral horns. Afferent nerves enter the cord on the posterior side. The entry area is swollen because it contains the cell body; these swellings are called *ganglia*. These nerves then *synapse* (connect) with motor neurons, which exit the cord on the anterior side.

### Clinical applications

- Polio specifically attacks anterior motor neurons, causing paralysis.
- Afferent and efferent nerves run in separate groups called *tracts*. Surgeons can cut ascending tracts without affecting descending tracts to block pain signals.
- Inflammation of spinal joints (arthritis) causes pain and loss of function.
- Degeneration of the pads between vertebrae can cause the discs to herniate, causing pain and even paralysis.
- ALS (amyotrophic lateral sclerosis; i.e., "Lou Gehrig disease") causes progressive degeneration of motor neurons, eventually leading to respiratory paralysis.
- Multiple sclerosis causes nerves to harden sporadically, destroying nervous function. Because the damage occurs in random areas, diagnosis can be difficult.

### The reflex arc

The reflex arc is a survival mechanism that allows rapid response to changes in the environment. Because the brain is not involved, response time is shortened. The pathway for the reflex arc is as follows:

- The peripheral receptor (pain dendrite in the skin) receives intense heat when one places one's hand on a hot stovetop.

- The impulse travels up the nerve, through the neuron body in the posterior root of the spinal nerve, and into the dendrite of the posterior horn gray matter.

- The association neuron in the gray matter transmits the impulse to regions in the CNS (brain) and to the lower motor neuron, whose body is in the anterior horn.

- The impulse exits the cord in the lower motor neuron (anterior root) and signals the muscle to withdraw the hand from the stovetop.

- The reflex withdrawal takes place *before* the brain is aware of danger.

- A fraction of a second *later*, the brain perceives the pain— "Ouch!" The hand has already withdrawn, but the perception of pain lingers.

Most reflex arcs have no integrating neuron and are called *monosynaptic arcs*. The others are *polysynaptic*. Neurotransmitter chemicals cross the gaps in synapses to transmit nerve impulses.

Reflex arcs can be used for diagnosis to detect defects somewhere along the arc pathway.

- Tapping on the patellar tendon just below the knee tests the pathway from stretch receptors in the femoral muscles, to the cord, and back down to the femoral muscles.

- The basic reflex arc in the knee jerk is monosynaptic.

- A second part of the arc, however, includes a polysynaptic pathway that inhibits the antagonistic muscle (the hamstrings) so that the femoral muscle *can* move the leg.

- Such diseases as polio can cause hyperreflexia, which is an uncontrolled reflex.

### The spinal nerves

- The spinal nerves exit the spinal cord at segments to connect the body with the brain.

- The posterior and anterior roots unite distally to form a mixed nerve.

- The spinal nerves are densely surrounded by protective muscle and bone. This protective structure is a two-edged sword, because minor shifts can cause compression of nerves and cord.

**The doctor will sit there with a pen and touch you while you have your eyes closed, and you've got to tell where the pen is touching or not. And we can get a very accurate indication of what part of the brain ... [has] been interrupted.**

- The mixed nerves are covered with an *epineurium* and group into bundles called *fascicles*, which are covered with a *perineurium*. An *endoneurium* covers each individual axon.

### Clinical application

The endoneurium of a severed nerve can be sutured back together, and the end still connected to the body may regenerate the whole nerve. Nerves form networks similar to anastomoses called *plexuses*.

- The cervical plexus supplies the neck and diaphragm.

- The brachial plexus supplies the arms. Its large nerves and sensitive location make it easy to injure.

- The lumbar plexus supplies the legs.

- The sacral plexus supplies the pelvis, buttocks, and lower limbs.

Spinal vertebrae can shift and narrow the spinal column (spinal stenosis). ■

## Questions to Consider

1. What is the evolutionary purpose of the reflex arc? How might it aid in survival?

2. Describe the steps in the knee jerk (patellar tendon) reflex.

# Nervous System—
# Autonomic Nervous System and Cranial Nerves
## Lecture 10

**Today we're going to continue and wind up our general look at the nervous system by examining the anatomy and the physiology of the autonomic nervous system and then of all the cranial nerves.**

The autonomic nervous system is the set of nerves that is functioning all the time, whether you're awake, asleep; sometimes even when you're in coma. Within the autonomic nervous system , the parasympathetic nervous system promotes rest and recovery, and the sympathetic nervous system promotes "fight and flight." The two main pathways for the autonomic nervous system are the general visceral motor (efferent) neurons and the general visceral sensory (afferent) neurons.

The entire system is involuntary. Some degree of cerebral control is possible during times of extreme fear and anxiety. Modern biofeedback methods have achieved some control in trained individuals.

Here are some differences between the autonomic and somatic nervous systems:

**Somatic nervous system**

- Motor activity is always excitatory—muscles at rest are signaled to perform specific tasks.
- Sensory input is conscious.
- There is only one neuron between the CNS and a given effector organ, making for a very, very long cell in some cases.

**Autonomic nervous system**

- Motor activity is *either* excitatory or inhibitory.
- Sensory input is generally unconscious, although extremes of sensation may be consciously perceived.
- There are two neurons between the CNS and an effector organ, with a synapse in the paraspinal ganglion.

**Divisions of the autonomic nervous system**
The parasympathetic system promotes rest, recovery, and relaxation. It is generally an inhibitory pathway. It is mediated by acetylcholine at the end organ. It has a craniosacral origin in the CNS, and there is very little divergence (1:3) at the ganglia.

The sympathetic system promotes "fright, fight, and flight." It is generally an excitatory pathway. It is mediated by norepinephrine at the effector organ. It has a thoracolumbar origin in the CNS. There is extreme divergence (1:20) at the ganglia. The sympathetic system also stimulates adrenal glands, which excite effector organs in the entire body.

**The sympathetic and parasympathetic nervous systems**
The sympathetic system prepares the body for fright, fight, and flight, which has the following effects:

- The pupils dilate.
- The heart rate and force of contraction increase.
- Blood glucose levels rise.
- Respiratory effort increases in rate and depth.
- Blood is shunted away from nonessential organs.

- Blood is shunted to the skeletal muscles, the myocardium, and the liver (for making glucose and fatty acids).

- The GI tract is slowed down or stopped.

- The adrenal medulla is stimulated to release epinephrine and norepinephrine (norepinephrine is the chemical mediator at the junction between the somatic nervous system axon and the target organ).

Accidental injection of adrenaline (epinephrine) into the bloodstream acts on the somatic nervous system and causes not only physical excitation but also extreme terror in the patient.

The parasympathetic nervous system prepares the body for rest, recovery, and relaxation, and it promotes energy conservation. When activated:

- The pupils constrict.

- The heart rate and force of contraction decrease.

- Blood glucose levels fall.

- Respiratory effort decreases in rate and depth.

- The GI tract resumes normal function, ready for digestion and absorption.

Paradoxical parasympathetic fear reactions include involuntary defecation and urination.

**The cranial nerves**

There are 12 pairs of cranial nerves. Ten of these pairs originate in the brain stem and exit the cranium through openings called *foramina* (singular: *foramen*). Mixed nerves contain sensory and motor fibers. Sensory nerves contain only sensory fibers. Both visceral and somatic motor fibers are represented. Motor neurons originate within the brain itself.

- The *olfactory (I) nerve* is sensory. It controls the sense of smell.

- The *optic (II) nerve* is sensory. It sends visual information from the eye to the brain.

- The *oculomotor (III) nerve* is mixed. This nerve, along with nerves IV and VI, controls the eye.

    - Its motor functions include control of the eyelids, the extrinsic muscles of eye movement, the sphincter of the iris (parasympathetic), and the ciliary muscle of the eye.

    - Its sensory function is to control the eye proprioceptors.

- The *trochlear (IV) nerve* is mixed.

    - The motor nerve controls the superior oblique muscle of the eye, causing downward movement in medial gaze.

    - The sensory nerve controls eye proprioception.

- The *trigeminal (V) nerve* is mixed. Its motor function is to control the muscles of mastication (chewing). It has three branches, each of which is mixed.

    - The ophthalmic branch controls the eyelids and forehead, eyeballs (cornea sensation), lacrimal glands, and the skin of the nose, anterior scalp, and forehead.

    - The maxillary branch controls the inside of the nose, the palate, the upper teeth, the upper lip, and the lower eyelids.

    - The mandibular branch controls the anterior two-thirds of the tongue (somatic, not taste), the lower teeth, the cheek, and the side of the head.

- The *abducens (VI) nerve* is mixed.
    - The motor nerve controls the lateral rectus muscle of the eye.
    - The sensory nerve controls eye proprioception.
- The *facial (VII) nerve* is mixed.
    - The motor nerve controls the facial (expression) muscles and the parasympathetic fibers to the lacrimal ducts, as well as the sublingual, submandibular, and nasal glands.
    - The sensory nerve controls taste from the anterior two-thirds of the tongue and face and scalp proprioception.
- *The vestibulocochlear (VIII) nerve* is sensory. The vestibular branch controls balance and the cochlear branch controls hearing.
- *The vagus nerve (X)* is mixed.
    - The motor nerve controls the muscles of the airway, the vocal apparatus, the lungs, the heart, and the entire GI tract. It also controls the parasympathetic fibers to the muscles and glands of the GI tract.
    - The sensory nerve has the same distribution as the motor nerve.
- The *glossopharyngeal (IX) nerve* is mixed.
    - The motor nerve controls the stylopharyngeal muscle (swallowing) and the parasympathetic fibers to the parotid gland.
    - The sensory nerve controls the taste buds from the posterior third of the tongue and the proprioceptors of swallowing.

- The *spinal accessory (XI) nerve* is mixed.
    - The cranial portion of the motor nerve controls the pharynx, larynx, and soft palate.
    - The spinal portion controls the trapezius and sternocleidomastoid muscles.
    - The sensory nerve controls the proprioceptors of the above muscles.
- The *hypoglossal nerve (XII)* is mixed.
    - The motor nerve controls the tongue.
    - The sensory nerve controls tongue proprioception.

**Clinical applications**
Fracture of the cribriform plate leads to loss of smell (*anosmia*) and cerebrospinal fluid (CSF) *rhinorrhea* ("leakage"), which can lead to brain infection.

Lesions along the oculomotor nerve can lead to *diplopia* (double vision) and *ptosis* (eyelid droop). Lesions of the abducens nerve cause medial deviation of the eye.

Dysfunction of the trigeminal nerve can lead to paralysis of chewing muscles and to *tic douloureux*—lancinating pain in the face, resulting from arterial compression of the nerve. Bell's palsy (paralysis of half the face), can result from either surgical or viral damage to the facial nerve.

*Tinnitus* (ringing in the ears), *vertigo* (dizziness), and loss of balance can develop from infections of the vestibulocochlear nerve.

Damage to the glossopharyngeal nerve causes *dysphagia* (difficulty in swallowing). Injury to the recurrent laryngeal nerve to the larynx (perhaps caused by surgical error during neck surgery) causes hoarseness.

Damage to the spinal accessory nerve (i.e., from a surgical error during neck surgery) causes shoulder droop.

Injury to the hypoglossal nerve causes difficulty in swallowing; tongue deviation points to the side with a lesion.

Hematoma or other pressure on the brain forces the brain down and can trap cranial nerves between the brain stem and the foramen magnum. ■

## Questions to Consider

1. How does sympathetic stimulation prepare an individual for fight?

2. What is the effect of parasympathetic stimulation on the cardiac muscles? On respiration?

# Nervous System—The Eyes

## Lecture 11

**Today we're going to examine one of the most extraordinary of the special senses, vision and the human eye. Now the human eye has a great range of capabilities. It may not be able to see as much in the dark as an owl can, but it's a very adaptable and another extraordinary structure.**

We begin our study of the eye by examining the anatomy of the eyeball—its various coverings and their functions (the cornea, sclera, uvea, and retina), and the photoreceptors of the retina ("rods" and "cones") that allow us to perceive different shades and colors of light. We also review the structure and functions of other components of the eye (the lens, eyelids, lacrimal glands and ducts, and extrinsic eye muscles). Next, we consider how the eye perceives light and how the brain converts those perceptions into meaningful information. Finally, the lecture reviews the most common eye disorders and their treatments.

Human eyes are not the most advanced in nature, but binocular vision gives them the ability to judge distances accurately. Loss of vision in one eye destroys this depth perception.

### The anatomy of the globe (eyeball)

The eyeball has several coverings. The *conjunctiva* is a thin, transparent layer that covers the eyeball. It is continuous with the eyelids.

The cornea is the outermost fibrous layer. It covers the iris and is transparent and nonvascular. It can help to focus light because it is curved. It is the most frequently transplanted tissue, because it is readily available from cadavers, and avascularity prevents circulating antibodies from attacking the transplant. It contains only pain fibers; thus, any contact with the cornea is interpreted as pain.

The *sclera* (meaning "hard") is the white part of the eye. It is a firm covering that maintains the shape of the globe. It provides some protection for the inner structures of the eye.

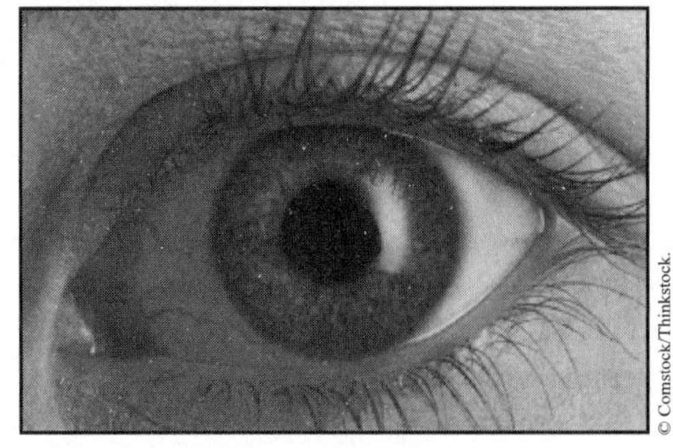


**The sclera is the white of the eye, and the iris is the colored portion.**

The *uvea* is the middle (vascular) layer. It provides most of the vascularity to the eye's coverings. The *iris* is the colored portion of the uvea. It contains circular muscles (parasympathetic) that constrict the pupil and radial muscles (sympathetic) that open the pupil. Legal and illegal drugs can mimic the parasympathetic system and close the iris.

The *ciliary body* consists of the ciliary muscle, which changes the shape of the lens, and the ciliary processes, which secrete the fluid that fills the eye.

The *choroid* provides the vascular supply to the retina. The retina is the inner (nervous) layer at the back of the eye.

- The visual surface of the retina is where the perception of light begins.

- The optic disc contains the blind spot, the egress of the central retinal artery and central retinal vein, and the egress of the optic nerve (CN II).

The retina contains 120 million rods—that is, rod-shaped photoreceptors. They allow one to see shades of gray in dim light and to perceive movement and shapes. Rods increase in number as one moves peripherally (to the outside, or temporal side, of the eye). Therefore, the best vision in darkness is peripheral, a survival advantage in avoiding predators.

The retina also contains 6 million cone-shaped photoreceptors. The cones allow for color vision in bright-light conditions. No colors are seen in dim light levels (e.g., night vision), because the cones do not function in dark conditions.

The *macula lutea* ("yellow spot") is at the direct center of the retina. The *central fovea* is a depression that contains only cones and is bared of coverings. Therefore, it is the part of the retina that provides the clearest vision.

The eyeball consists of three chambers. The anterior chamber is between the cornea and the iris. The posterior chamber is between the iris and the lens. The vitreous chamber is between the lens and the retina; it is filled with vitreous, a jelly-like substance.

**Clinical application**
Many diseases can be diagnosed by looking into the eyeball. Glaucoma can result when pressure builds up in the eyeball, either from overproduction of aqueous humor, which nourishes the lens, or failure of the canal of Schlemm, which regulates intraocular pressure. Retinal tears expand blind spots; retinal detachment distorts vision.

**Accessory structures of the eye**

- The *palpebrae* (eyelids) spread tears (acting as "windshield wipers") and protect the eye from trauma.

- The eyelashes filter dust and debris and signal their approach.

- The lacrimal apparatus consists the lacrimal gland, which contains a lysozyme, and the superior and inferior lacrimal ducts. The lacrimal glands generate 1 ml of fluid per day, which normally drains into the nose. Crying and irritation increase fluid flow.

- There are six extrinsic eye muscles in each orbit:

    - The lateral and medial rectus muscles control lateral (to the outside) and medial (toward the body's midline) movements.

- The inferior and superior obliques control rotational, as well as upward and downward movement. The eyes can rotate slightly to keep perception of the horizon relatively flat.

- The superior and inferior rectus muscles control upward and downward movements.

To suppress blurring, the brain directs the eyes to pick a fixed point in a moving landscape, follow it until it disappears, then pick another point. This is called *nystagmus*. Abnormal nystagmus can indicate brain damage.

The six muscles are supplied by cranial nerves III, IV, and VI. They provide the finest motor control in the body.

**Physiology of light and vision**

The visual pathway through the brain is as follows:

- The optic nerves cross at the optic chiasm and run into the brain's *visual cortex* in the occipital lobe.
    - The outer fibers remain ipsilateral (no crossover).
    - The inner fibers cross over to the contralateral side.
    - Chiasm fibers end in the lateral geniculate body.
    - The visual cortex reorganizes visual objects—their shape, movement, size, color, and orientation in space—into the picture we see.
- The photoreceptors function in the following manner:
    - Light strikes the retina, stimulating the rods and cones.
    - Photoreceptors convert light energy into chemical energy.
    - Chemical energy is converted into nerve conduction.

- Chemical action is Vitamin A–dependent for renewal.

- The signal finally reaches the neurons in the optic nerve (cranial nerve II).
  - The rods tend to converge into one nerve cell and thus offer better vision in dim light, at the cost of blurring.
  - The cones tend to have one nerve each and, therefore, give sharper images in color while sacrificing sensitivity.

**What can go wrong?**
Glaucoma

- Increased intraocular pressure damages the retina.
- It is painless and insidious, except for acute cases.
- Surgical and medical treatment is available for glaucoma.

Retinal detachment

- Retinal detachment usually results from trauma or acceleration/deceleration.
- The superficial layers are separated.
- Surgical, laser, and cryogenic procedures are available for correction.

Cataracts

- Cataracts are the leading cause of loss of vision.
- They are opacities in the lens and cause vision clouding.

- They result from injury, UV light, aging, diabetes, smoking, or steroids.

- They can be corrected through surgery, with or without implants.

*Infections and trauma*

- Corneal injury.

- Conjunctivitis.

- Herpetic ulcers. ■

## Questions to Consider

1. What does the presence of more rods at the periphery of the visual field do to acuity of vision in that area?

2. Discuss convergence and divergence as they refer to the circuits of the rods and cones. Which confers greater visual acuity? Sensitivity?

# Nervous System—The Ears, Hearing, and Equilibrium

## Lecture 12

**In this lecture, we're going to cover another organ of special sense, that of the ear, which contains both the function of hearing and the function of balance. We do have to admit that this organ does not stand up to the high evolutionary standards of the eyes or other comparisons with lower forms of animals, other primates.**

We begin our study of the ear by examining the anatomy of the organs of hearing: the external ear; the "eardrum"; the tympanic cavity, or "middle ear"; and the labyrinth, or "inner ear." Next, we review how these structures gather and transmit sound waves to the eighth cranial nerve, which transmits them to the brain as nerve impulses. Finally, we examine the anatomy and functions of the vestibular apparatus, a set of structures in the inner ear that governs balance.

**Anatomy and physiology of the organs of hearing**

The external ear is designed to gather sound into the ear.

- The *helix* is the flared end of the ear. It is the edge of the *auricle* (the entire ear).
- The *lobule* is the ear lobe.
- The external auditory canal (*meatus*) is lined with hair cells and glands that produce wax (*cerumen*), which protects the ear from dirt.

Sound waves (air- or water-borne vibrations) are gathered by the auricle and funneled into the external auditory canal, where they strike the tympanic membrane, also known as the eardrum.

- It separates the external ear from the middle ear.

- It is taut, thin, and made of connective tissue between layers of epithelium.

- It moves back and forth at the pitch (rate of vibration) and amplitude (intensity of vibration) of incoming sound.

The eustachian tube connects the middle ear to the nasopharnyx (nose and mouth cavity) to equalize pressure. It is more effective at releasing air from the inner ear than at drawing air into the inner ear. Pressure differences between the inner and outer ear cause pressure, clogging, and pain.

**The deaf person, while he's able to see, drive, and move about with more ease than someone who is blind, has difficulty in social communication and the emotional issues involved there tend to be very difficult to deal with.**

The middle ear is called the tympanic cavity.

The auditory ossicles are three small bones that amplify vibrations received by the eardrum. They comprise:

- The *malleus* ("hammer"), which attaches directly to the eardrum.

- The *incus* ("anvil"), which is located between the malleus and stapes.

- The *stapes* ("stirrup"), which is located between the incus and the oval window. It is the smallest bone in the body.

Movement of the tympanic membrane causes movement of the malleus-incus-stapes bones in sequence. Mechanical advantage (from the bone shapes, their leverage, and the difference in large surface area between the tympanic membrane and the oval window) amplifies tympanic membrane vibration more than 20-fold.

The tympanic cavity is buried within the temporal bone. Two muscles limit the motion of the eardrum and limit damage from prolonged loud noises:

- The *tensor tympani* muscle attaches to the malleus and increases tension on the eardrum.

- The *stapedius* muscle attaches to the stapes and limits its motion. It is the smallest muscle in the body.

These muscles cannot react immediately and, thus, provide no protection against sudden loud noise. Nerve damage causes hyperacusis (heightened hearing) as these muscles become paralyzed.

The stapes is attached to the oval window, an opening in the *cochlea* ("snail shell"), the actual organ of hearing. It is y-shaped on cross-section and consists of a partitioned spiral containing three chambers:

- The *scala vestibuli* ends at the oval window.

- The *scala tympani* ends at the round window.

    - These scalae are filled with perilymph fluid, which enables the transmission of vibrations through the cochlea.

- The *cochlear duct* is located between the two scalae.

The spiral *organ of Corti* detects sound waves. It is filled with endolymph fluid and has multiple rows of microcilia (small hairs) that wave with the motion of the endolymph fluid and transmit those waves to the cochlear nerve. Oval window movement pushes fluid from the scala vestibuli to the scala tympani and, finally, to the round window. The cochlear duct moves in sync with the waves in the two scala.

**Clinical application**
Loud noise damages the ossicles, which can be corrected, and causes nerve deafness, which is irreversible. Earphones are particularly dangerous, because they concentrate noise on the ear. Cochlear implants directly stimulate the

hearing nerves. Deafness is a very difficult defect to adjust to, perhaps more so than blindness, because it isolates the patient from social interaction.

**Anatomy and physiology of the organs of balance (equilibrium)**
The *utricle* and *saccule* organs detect static equilibrium (orientation of the body) and dynamic equilibrium (changes in motion). They cannot detect uniform motion (that is, constant speed and direction).

- Motor and sensory nerves travel from the vestibular branch of the cochlea to the utricle and saccule.

- Cilia in both organs protrude upward into divided, jellylike otolithic membranes.

- *Otoliths* (calcified ear stones) rest on the membranes, moving with the body and stimulating the cilia.

- Stimulation of the cilia provides information to the brain about the body's orientation and motion.

Three semicircular canals, at right angles to each other, detect dynamic equilibrium. One canal is horizontal, one is vertical, and one travels from the anterior to the posterior sides of the body. This provides three axes to detect motion in any direction.

The canals are filled with endolymph fluid and surrounded by perilymph fluid. Each canal ends in a swelling (*ampulla*) connected to a *crista*, which sends nerve fibers to the vestibulocochlear nerve. Motion of the body sets the endolymph fluid in motion, stimulating cilia in the canals. Motion of the cilia produces nerve impulses that are sent to the brain.

**Clinical application**
Differences between visual cues and the motion of fluid in the canals cause disorientation and nausea, as in seasickness.

**Potential problems with hearing and balance**

- Perforation of the eardrum can cause scarring and decreased sensitivity.
- Exposure to foreign proteins (such as cow's milk) at a young age can cause middle-ear infections that require implantation of tubes.
- Patients with Meniere's disease produce too much fluid in the vestibular apparatus, causing continual motion of the cilia and disorientation, dizziness, and nausea.
- Nystagmus (rapid eye movement) in a person at rest signals a problem of coordination between the eye and the inner ear.

The brain coordinates information from the vestibular apparatus with the organs of proprioception to form a complete picture of the body's orientation and motion.

**Clinical application**
Patients with tertiary syphilis lose proprioception in their feet, causing them to walk with a very distinctive gait. ■

## Questions to Consider

1. Describe the components of the middle ear.

2. How is balance transmitted from the semicircular canals to the brain? Is this static or dynamic equilibrium?

# Nervous System—Memory

## Lecture 13

**Today we're going to finish up our section on the nervous system by looking at all the various things that can go wrong with segments of the system. We're going to divide it up basically by categories and see exactly what can happen to the nervous system to cause certain illnesses or diseases and malfunctions.**

In this lecture, we complete our examination of the nervous system with a look at memory and brain pathology and provide some information about anesthesia and pain. First, we examine the nature, development, pathology, and mysteries of memory. We then examine various kinds of damage to the brain and spinal cord and the results of such damage. We conclude by discussing anesthesia and referred pain, in which organs that share nerves with other body parts can cause the brain to report pain in those body parts.

Memory prevents repeated mistakes and promotes repeated successes. It has been extensively studied, but relatively little is known about it compared to other systems of the body. More is known about memory acquisition than about memory retrieval. *Learning* is the acquisition of new skills or information, either through experience or instruction. *Memory* is the process of storing that information and retrieving it when needed. The brain has high plasticity, which is the ability to change learned behavior on receipt of new information. Anatomically, memory involves changing the interconnection of neurons by creating, modifying, or destroying dendrites and synapses. The brain has almost all its neurons from birth, and it does not expand with new memories.

### Memory areas of the brain

- The entire brain is involved in memory processes.
- Different areas of the brain (the hippocampus, amigdala, diencephalon, and so on) are primary memory centers for certain functions.

- Centers of the brain devoted to controlling body parts will expand as those parts are used repeatedly.
- The brain stores only about 1% of conscious information in long-term memory, probably because of space limitations.

**Types of memory**

- Immediate memory is a few seconds long and consists of activities occurring at the present moment.
- Short-term memory lasts between a few seconds and a few minutes.

The hippocampus, mamillary bodies, and thalamus handle both immediate and short-term memory. Immediate and short-term memory are probably chemical and electrical events rather than anatomical changes involving dendrites and synapses. Repetitive use of information in short-term memory converts that information into long-term memory, which can last from hours to a lifetime.

**Memory storage**

- Verbal or written information is stored across the entire cerebral cortex. Individual ideas and memories cannot be excised or ablated.
- Physical activities are stored in the areas corresponding to the body parts used and in the basal ganglia.
- Brain scans show an increase in the number and size of presynaptic terminals in corresponding areas when information is retrieved repeatedly and a decrease in number and size when information is infrequently used.

**Possible causes of memory loss**

- Normal, proper use of anesthesia.
- Coma.
- Electroconvulsive shock therapy.
- Loss of blood flow to the brain (stroke, suffocation, etc.).
- Brain or head trauma.

**Types of memory loss (amnesia)**

- Circumscribed (retrograde) amnesia causes loss of memory directly before the traumatic event. If the memory returns, the oldest memories will return first and be followed by newer ones.
- Complete amnesia causes loss of all memory. It is very rare.

Animal studies are much less helpful for memory studies than for other subjects, because human memory appears to be far more advanced than animal memory.

**Pathology of the brain**

Cerebellum

- The cerebellum is the center of coordination.
- Damage from trauma, alcoholism, or ischemia (lack of blood flow) causes cerebellar ataxia, which causes difficulty in walking, speaking, and other highly coordinated functions.
- Intention tremor is a series of wild over-corrections when attempting to, for example, touch one's nose, in cases of cerebella ataxia.

Basal ganglia and Parkinson's disease

- The basal ganglia are involved in coordinated body functions but in a different way than the cerebellum.
- In Parkinson's disease, the *substantia nigra* ("black matter") releases dopamine, a neurotransmitter, to the basal ganglia.
- This dopamine release results in lowered dopamine levels and increased acetylcholine levels in the basal ganglia.
- Parkinson's has a specific set of symptoms, including waxy paralysis, which is overactivity of motor nerves, resulting in a limited range of movement and facial activity, including the typical "pill rolling" movement.
- Parkinson's is hard to treat because of the blood-brain barrier, which prevents direct delivery of dopamine. It is also progressive, and after a certain point, dopamine is ineffective.
- Direct destruction of brain tissue can prevent certain tremors, but no overall treatment has been developed.
- Parkinson's does not affect higher brain function, unlike Alzheimer's disease.

Huntington's chorea

- Huntington's chorea is a fatal disease from an inherited, genetically dominant gene that is easily transmitted.
- It is not selected against by evolution, because the disease does not develop until the patient has reached a mature age (typically 30 years old or more).
- *Chorea* means "dance"; patients have rapid, uncontrollable movements and progressive mental deterioration.

- Genetic testing can identify the Huntington's gene in prospective parents.

## Pathology of the spinal cord

- Causes of spinal cord damage include:
    - Trauma to the cord itself.
    - Injuries to the spinal column (e.g., herniated discs).
    - Tumors.
    - Blood clots.
- *Mono-* or *diplegia* is the paralysis of one or two extremities; *mono-* or *diparesis* is weakness in one or two extremities. Both are caused by compression or interruption of a nerve or nerves from the spine. If two extremities are affected, they are almost always on the same side.
- *Paraplegia* ("parallel limb paralysis") is a diplegia on both sides of the body, usually below the waist.
- *Quadriplegia* is paralysis of all four limbs.
- *Hemiplegia* is paralysis of one side of the body and is usually a problem in the brain (stroke, tumor, etc.).
- Brown-Sequard syndrome is caused when one side of the spinal cord is damaged but not the other.
    - The patient will have paralysis or weakness below the injury on the same side.
    - The patient will lose temperature and pain sensation below the injury on the opposite side.

- If the nerves controlling respiration are cut, the patient cannot breathe without assistance.

- Spinal injury can cause spinal shock, where the spine itself goes into shock. Symptoms include:

    - A drop in blood pressure, because capillary bed constriction is uncontrolled.

    - Loss of reflex (*areflexia*); even if the reflex arcs are intact, the spine will not respond.

    - Flaccid paralysis, including loss of limb and bowel control.

    - Reflexes and muscle control will return if the shock is halted.

**Anesthesia**

A patient under general anesthesia is asleep; there is no reception of pain sensation even though peripheral receptors are sending messages. Under local anesthesia, peripheral nerves are blocked chemically, and no pain impulse reaches the brain. The patient is awake.

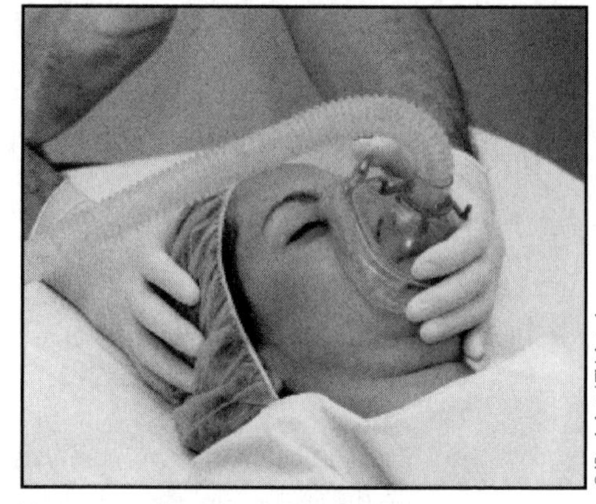



**Under anesthesia, peripheral receptors send pain messages, but the brain does not receive them.**

**Referred pain**

Sensory nerves from visceral organs are carried in the same fibers to the brain as other skin and muscle nerves. Organ damage can cause pain in specific areas of the body where nerves share pathways, because the brain cannot tell the difference between impulses.

Examples of referred pain include:

- Heart damage causes pain in the left arm.
- A ruptured spleen causes left shoulder pain.
- Appendicitis causes pain in the lower abdomen. ■

## Questions to Consider

1. What is the anatomic basis of the differences between short-term and long-term memory?

2. Describe the differences between movements in patients with Parkinson's disease (rest tremor) and cerebellar ataxia (intention tremor).

# Digestive System—Anatomy of the Mouth, Esophagus, and Stomach

## Lecture 14

**Today, we're going to begin a new topic: The digestive system. And we're going to start with the anatomy and, again, work slowly into the physiology of the digestive system in sections. We're going to start with the upper digestive system and work our way down.**

This is the first lecture in a six-lecture examination of the digestive system. We begin by examining the anatomy of the structures through which food passes before its conversion into nutrients for the body. We conclude by reviewing the four main functional divisions of the stomach (the cardia, fundus, body, and antrum) and its various layers.

The digestive system is also known as the gastrointestinal (GI) tract and the alimentary canal. It is a continuous tube beginning at the lips and ending at the anus. Embryologically, it forms from a tube-like structure that develops outpouchings, which in turn evolve into attached digestive organs, such as the pancreas, liver, gallbladder, and so on. The entire system is about 40 feet in length; it is designed to transport food and water, modify it, and make it suitable for absorption and excretion. There are storage sites, excretion sites, and detoxifying sites along the way.

Food enters the body through the mouth, where it is moistened by saliva for ease of swallowing. The tongue then pushes the food into the esophagus, which carries the food to the stomach. Let's examine this process and the structures that participate in it in some more detail.

**Lips and mouth**

The lips function in eating and forming sound.

The mouth is lined with mucous membranes that lubricate its surfaces. The teeth cut and grind food; cutting teeth are in the front of the mouth, with the grinding teeth in the back. The roof of the mouth is the hard palate anteriorly and the soft palate posteriorly.

The rear of the mouth is defined by the soft tissue arch separating the mouth from the throat. It has two parts: the palatoglossal arch and the palatopharyngeal arch. Tonsils between the two arches fight infection. The uvula hangs down from the arch in the midline. It rises during swallowing to close off the nose from food.

The tongue is the largest muscle mass in the mouth. It functions in chewing, swallowing, and forming words.

- Extrinsic muscles of the tongue attach to the skull and neck, and they move from side to side and in and out.
- Intrinsic muscles attach to the tongue itself, and they alter the tongue's shape (for swallowing and speech).
- *Papillae* are the bumps on the tongue that contain taste buds.

There are three pairs of salivary glands in the mouth. They secrete saliva, the first of the digestive juices to contact the *bolus* (a lump or wad of food). The three pairs of glands are:

- *Parotid* ("near the ear") glands.
    - These glands are under the skin on top of the masseter muscle.
    - A duct passes around the masseter to enter the mouth between the molars.
    - Mumps is a viral infection that affects the parotid glands (as well as the testes and ovaries).
- Submandibular glands.
    - These glands are on the floor of the mouth, just under the mandibles.
    - They enter the floor on either side of the frenulum of the tongue.

- Sublingual glands.
  - These glands are on the floor of the mouth.
  - Ducts enter beneath the tongue.

The saliva moistens the mucous membranes, moistens food for easy swallowing, dissolves food, aiding taste perception; lubricates the esophagus for swallowing, washes the mouth, kills bacteria, and dilutes poisonous substances. (A surgical maxim states: "The solution to pollution is dilution.")

- The output of saliva is 1–1.5 liters per day.
- More than 99% of saliva is water, most reabsorbed in the GI tract.
- Saliva also consists of 0.5% enzymes.
- Salivary amylase begins the breakdown of carbohydrates.
- Lysozyme kills bacteria in the mouth.
- Ions include bicarbonate (buffers acid), chloride (activates amylase), and phosphate (part of the buffer system).
- Lingual lipase breaks triglycerides down into fatty acids and a monoglyceride.

During dehydration, the brain signals the mouth to stop the flow of saliva to impel us to drink more and to conserve fluids. Pavlov's dogs salivated to the sound of a bell that they associated, from prior training, with food.

**Physiology of swallowing (*deglutition*)**
Swallowing is divided into three stages:

- During the voluntary stage, the tongue moves the bolus of food back by its upward and backward movement.

- During the pharyngeal stage (involuntary), the muscles move the food down and back into the esophagus.

- During the esophageal stage (involuntary), the food is moved through the esophagus to the stomach by a series of constrictor muscles acting in a coordinated fashion.

Aspiration (entry of food or water) into the lungs and nasopharynx is prevented in the following way:

- The uvula and soft palate move upward to close off the nasopharynx.

- The larynx is pulled forward and upward under the protection of the tongue.

- The epiglottis moves back and down to close the opening of the trachea and airway.

- Food slides over the epiglottis into the esophagus.

- Vocal cords close to further block the airway.

- Breathing ceases for about 2 seconds while this process takes place, then resumes.

**Esophagus ("carries food")**
The only function of the esophagus is to carry food from the mouth to the stomach. No absorption occurs here.

The passage of food through the esophagus is active, not gravity-driven. Longitudinal muscles pull the esophagus up and relax lower portions so that circular bands of muscle can constrict and move the bolus down into the stomach. One *can* swallow while inverted, but this is not recommended.

The esophagus traverses the mediastinum, directly in front of the spinal column and behind the trachea in the neck and upper chest. It has no *serosa* (outer covering), unlike the other digestive organs. Water goes through quickly (in 1 second) and food may take longer (5–9 seconds).

The lower esophageal sphincter is located just above the diaphragm. It is physiologic, not anatomic. It allows food into the stomach while preventing reflux of stomach acids and bile into the esophagus.

The sphincter is the site of a good deal of pathology, including:

- *Achalasia* (inability to relax), which prevents food from entering the stomach.

- *Gastroesophageal reflux disease* (GERD), also known as acid reflux, in which the sphincter fails to prevent acid from backing up into the esophagus. This causes inflammation, scarring, and sometimes cancer.

- *Hiatal hernia*, in which the stomach moves above the diaphragm and into the chest.

New drugs have radically improved the treatment of hiatal hernia and GERD.

**Stomach**

The stomach is an enlargement and rotation of the GI tube, just below the diaphragm. The rotation affects the placement of nerves along the outside of the stomach. Its functions are to mix and digest. It is a massive enlargement compared to the esophagus.

The single cavity has four main functional divisions:

- The *cardia* is a small space just under the diaphragm, under the heart.

- The *fundus* is the main upper portion. It reaches to the top left of the esophagus.

- The *body* is the large middle section.

- The *antrum* ("cave") is the last part of the stomach before the *pylorus* ("gatekeeper"), which prevents food from entering the intestine before it or the food are ready.

Food moves sequentially through these sections, rather than just dumping into one great cavity. We are able to consume more than the intestines are ready for at one time. The divisions allow us to process the food slowly and prepare it for entry into the intestines.

The stomach has the following layers:

- The *serosa* (also called the *visceral peritoneum*) carries blood vessels and protects the stomach.

- Longitudinal, circular, and oblique muscle layers and *rugae* ("folds") on the stomach's interior provide grinding and mixing motions.

- The *muscularis mucosa* moves and supports the mucosa.

- The mucosa is a thick, plush layer of cells that lines the stomach cavity. It has deep clefts that increase the stomach's surface area considerably. Only water and alcohol are absorbed through the mucosa. Cells of the mucosa include the following:

    - Mucosal neck cells are the most numerous cells in the stomach. They produce a thick glycoprotein that protects the stomach wall from autodigestion.

    - Parietal cells produce stomach acid (HCl) and *intrinsic factor*, which helps absorb vitamin B-12. They are located only in the fundus and the body.

    - Chief cells produce pepsinogen (precursor to pepsin) and gastric lipase (a fat-digesting enzyme).

    - Enteroendocrine cells (G-cells) in the antrum produce the hormone gastrin, which is secreted directly into the bloodstream to stimulate the parietal cells to produce hydrochloric acid.

Because humans are predators, as well as prey, the human digestive tract has evolved to take in food in large amounts with long intervals between meals.

The blood supply to the stomach is rich and redundant.

- The *omentum* is a thick organ with many blood vessels that connects the stomach to the lower abdomen. The omentum changes from slick to sticky with abdominal disease or trauma.
- The celiac trunk is the first abdominal branch off the aorta. It branches into the following arteries that serve the stomach:
  - The common hepatic artery branches into the gastroduodenal and left and right gastric arteries.
  - The splenic artery branches into the left gastroepiploic and short gastric arteries.
- Many of these arteries and subarteries anastomize to provide redundant blood supplies to the stomach.
- The stomach is difficult to devascularize, which can be helpful or fatal during trauma and surgery. ■

## Questions to Consider

1. Detail the safeguards that prevent food from entering the airway during swallowing.
2. How is gastrin different from the rest of the gastric secretions listed in this lecture?

# Digestive System—Physiology of the Mouth, Esophagus, and Stomach

## Lecture 15

**Today we're going to pick up where we left off in anatomy and now follow food as it goes from the mouth to the antrum of the stomach and see the physiology and the chemical processes that are at work, just before we move the food into the duodenum and the next portion of the intestinal tract.**

Having studied the anatomy of the mouth, esophagus, and stomach, we turn now to the processes of digestion, absorption of nutrients, and excretion of waste products. This lecture examines, first, mechanical digestion (the stomach's mixing of the food, now called *chyme*), then chemical digestion (the action of various acids, hormones, and enzymes on the chyme). Next, we examine the three phases of gastric secretion and the process of gastric emptying, as the chyme passes from the stomach to the duodenum. Finally, we consider various stomach and digestive disorders and their treatments.

We have already reviewed the mouth and esophagus. Breakdown of carbohydrates and some triglycerides begins in the mouth but is not completed there. The esophagus is merely an active passageway to the stomach (designed not to take up too much room in the chest). We proceed now with a bolus of food arriving in the cardia of the stomach.

### The processes

- *Digestion* is the physical and chemical breakdown of nutrients into particles suitable for absorption.
- *Absorption* is the movement of these broken-down nutrients from the digestive tract into the bloodstream for use by the cells of the body.
- *Excretion* is the expulsion of the unused products of digestion and absorption from the body.

- During *detoxification*, ingested material that might otherwise be toxic is rendered harmless.

- Summary of the process to this point:

  - The mouth begins digestion but has no role in absorption or excretion.

  - The esophagus is a conduit and has no role in absorption or excretion.

  - The stomach is an organ of digestion and limited absorption but not of excretion.

  - None of these is involved in detoxification; the primary detoxifier is the liver.

**Two primary digestion processes in the stomach**

Mechanical digestion

- Waves of *peristalsis* (muscle contractions) grind and mix the food.

- The bolus is mixed with stomach juices and is now called chyme.

- Very little mixing takes place in the fundus.

- Salivary enzymes continue to work in the fundus, because they have not yet mixed with acid gastric juices.

- In the body of the stomach, the bolus of chyme is exposed to and mixes with gastric acid, and salivary enzymes are inactivated.

- More grinding and mixing occur in the body of the stomach.

- At the antrum, there is a purposeful holdup. Only a small amount of chyme can get into the duodenum at a time; the rest remains in the antrum for additional mixing and grinding.

Chemical digestion

- The parietal cells secrete hydrochloric acid (HCl) in very high concentrations. HCl performs the following functions:
    - It denatures (unfolds) proteins to prepare for digestion.
    - It kills many micro-organisms, such as the ones in the human mouth or in foreign matter.
    - It stimulates the flow of hormones, bile juices, and pancreatic juices in preparation for release into the small intestine.
    - It inhibits the activity of the hormone gastrin in a negative feedback loop.
    - It stimulates the release of secretin and cholecystokinin (CCK) and prepares the small intestine.
- The chief cells secrete pepsinogen, which is converted into pepsin by HCl. It is active only at low (acidic) pH. Pepsin breaks bonds between the amino acids that make up protein; thus, it begins the digestion of protein.
- The chief cells also secrete gastric lipase, which breaks triglycerides into fatty acids and a monoglyceride. Because gastric lipase is active at lower pH than the stomach, its role is limited here.
- Lingual lipase and gastric lipase are overshadowed by the later action of pancreatic lipase.
- Mucous cells keep the stomach juices from digesting the stomach (sometimes!).

- Enteroendocrine cells (G-cells) stimulate the release of HCl and pepsinogen.
    - Pepsinogen increases gastric motility.
    - It relaxes the pyloric sphincter to enable food to get into the duodenum.
    - It closes the lower esophageal sphincter to prevent reflux.

Gastric secretion and control of motility go through three phases. These phases are regulated by both neural and hormonal factors. There is much overlap.

The first phase is the cephalic phase (neural). The cephalic phase is mediated by the parasympathetic vagus nerve (cranial nerve X). The vagus nerve controls muscular contraction, stimulates secretion of HCl and pepsinogen, and stimulates mucous production. It also stimulates the release of gastrin from the antrum. Pavlov's dogs' responses (as well as our own) are vagus-mediated.

- Cortical perception (thinking about food).
- Hypothalamus and medulla oblongata (brain stem).
- Vagus nerve.

Sympathetic stimulation reverses cephalic phase actions.

The second phase is the gastric phase (neural and humeral). This phase is initiated by the presence of the bolus in the stomach. (The presence of food is not required during the cephalic phase; the sight, smell, or thought of food is sufficient.) It is activated by distension (stretching) of the stomach wall, which is mediated by the vagus nerve. More gastric juice is secreted, and the cycle intensifies.

Food neutralizes the acid pH of the stomach. Chemoreceptors in the stomach detect the rise in pH. Enteroendocrine cells are stimulated to secrete gastrin into the bloodstream. Gastrin circulates to the stomach and stimulates HCl output from the parietal cells. Antacid tablets that neutralize acid in the stomach, therefore, actually stimulate the chemoreceptors to produce more acid. The total effect is to force a small amount of chyme across the pylorus into the duodenum. Distension decreases and pH decreases (increase in acidity). The negative feedback loop slows down the process. Gastrin shuts the esophageal sphincter and opens the pylorus.

The third phase is the intestinal phase. During the cephalic and gastric phases, the intestinal phase effects are inhibitory. The intestines are trying to slow down gastric digestion and emptying to give themselves more time to prepare for food. The functions of this phase include additional digestion, as well as the first significant absorption.

The duodenum is the first 12 inches of the small intestine, before it enters the colon.Three hormones are released by the entry of chyme (especially fatty acids and glucose) into the duodenum:

- The gastric inhibitory hormone (GIP) inhibits gastric secretion and motility.
- Secretin decreases gastric secretion.
- Cholecystokinin (CCK) inhibits gastric emptying.

The neural reflex is stimulated by distension (chyme) in the small intestine. It mediates the sympathetic nervous system and inhibits the parasympathetic nervous system. It results in gastric motility and secretory inhibition.

**Gastric emptying**

Gastric emptying is promoted by gastric distension, partially digested protein fragments (amino acids), and drugs (e.g., alcohol and caffeine). All of the above tend to increase gastrin secretion and vagal stimulation. All of the above tend to close the lower esophageal sphincter, open the pylorus, and increase gastric peristaltic contractions. Gastric emptying is inhibited by duodenal distension (by the chyme bolus), fatty acids, glucose, protein fragments, and increased secretion of CCK and GIP. The stomach empties more slowly.

Foods spend varying times in the stomach. Fatty foods remain the longest time. Proteins remain an intermediate length of time. Carbohydrates remain the least time. Combining all three at a meal gives the most effective satiety index (a measure of how full the stomach feels). Not only can the stomach stretch quite a bit, but it tends to collapse quickly when stretched, causing hunger quite soon after a large meal.

**What can go wrong?**

Alcohol is readily absorbed from the stomach (water can be absorbed as well). It is very quickly absorbed from the small intestine. Alcohol dehydrogenase breaks down alcohol (detoxification). Because females have much less of this enzyme, they are more readily affected by alcohol ingestion. Food in the stomach (especially fats) slows down alcohol absorption. Regular alcohol use will raise not only the amount of alcohol dehydrogenase but also the amount of other hormones, which can inhibit the effectiveness of medicines.

Vomiting is a process of reverse peristalsis to rid the body of toxic elements. It is secondary to gastric distension or distal obstruction. Blood is one of the most distressing fluids to have in the GI tract and causes immediate vomiting. Prolonged vomiting can cause fluid and electrolyte imbalance and tears in the esophageal veins (Mallory-Weiss syndrome). Extensive tears can rupture the esophagus and spill food into the mediastinum (medistinitus), which can be rapidly fatal.

Ulceration in acid-exposed areas is called peptic ulcer disease (PUD). It occurs in the first portion of the duodenum and stomach. Its causes include the following:

- *Helicobacter pylori* produces a urease, liberating ammonia, which neutralizes acid. Catalase protects bacteria from host defenses. This type of ulcer requires antibiotics to kill the bacteria, as well as treatment of the ulcer.

- Nonsteroidal anti-inflammatory drugs (NSAIDs).

- Other drugs, including caffeine, alcohol, and nicotine.

Old treatments for PUD involved major surgery to cut the vagus nerve to parietal acid-producing cells. Proton-pump inhibitor drugs have dramatically changed PUD treatment, because they block the stimulation of acid-producing cells without surgery. ■

## Questions to Consider

1. Describe the cephalic phase of gastric secretion. What is the mediator in this process?

2. Describe the intestinal phase of gastric secretion. Contrast it to the other two phases.

# Digestive System—
# Anatomy of the Pancreas, Liver, and the Biliary Tree
## Lecture 16



**Today we're going to continue on our way down the gastrointestinal tract and what we're going to do is focus on those out pouchings that I mentioned that occur during embryology and then develop into the liver, the pancreas, and what we call the biliary tree, which is the inside of the liver.**

This lecture examines the anatomy of the pancreas, liver, and gallbladder. First, we examine the gross and microscopic anatomy of the pancreas, a 12-inch-long organ that produces hormones and digestive juices and secretes them into the duodenum. We turn next to the liver, the body's largest and heaviest gland, and its second-largest organ (after the skin). The location, size, and blood supply and routing of the liver are reviewed. Finally, the lecture examines the gallbladder and biliary tree—the complicated duct system that drains bile from the liver into the gallbladder and the duodenum.

The peritoneal cavity is the abdominal space containing the intestines, stomach, and liver. The peritoneal cavity is lined with a parietal peritoneum, and the organs inside the cavity are covered with a visceral peritoneum, as in the pleura and the pericardium. Retroperitoneal organs lie within the peritoneal cavity but outside (behind) the visceral peritoneum. The pancreas, the kidneys, the middle of the duodenum, and the rectum are important examples.

### The pancreas

*Pancreas* means "all flesh," which refers to its ability to digest virtually all proteins, including itself. It is a retroperitoneal organ located in the posterior part of the middle to upper abdominal cavity. It is about 12 inches long, tapering from right to left. The head lies to the right, nestled in the duodenal sweep. The body traverses up and to the left, tapering to the tail. The tail terminates at the junction with the spleen.

The posterior wall of the stomach lies against the anterior surface of the pancreas. This is an important clinical relationship for cases of pseudocysts of the pancreas. The first portion of the duodenum lies in front of the pancreas. This is important in complications of duodenal ulcer disease.

The pancreas is richly supplied with arteries and veins. It is served by branches from the hepatic artery, the gastroduodenal artery, the pancreaticoduodenal artery, the superior mesenteric artery, and the splenic artery.

A duct system drains bile into the duodenum. There is a major pancreatic duct (the duct of Wirsung) and an accessory pancreatic duct (the duct of Santorini). The ducts meet at the right side to empty through the ampulla of Vater into the duodenum. The main duct usually joins the common bile duct before entering the duodenum. This is important clinically in common channel pancreatitis.

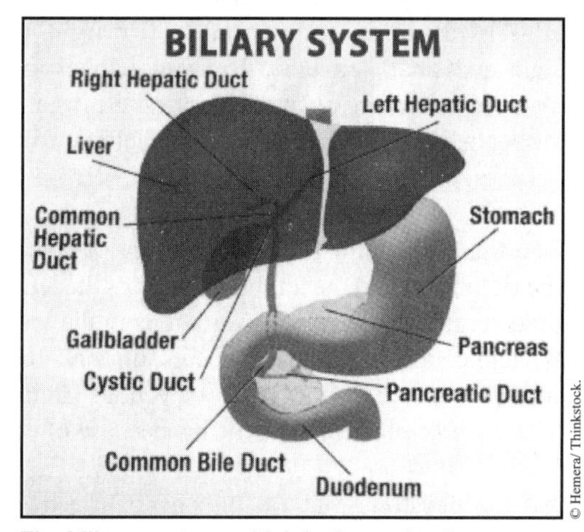


The biliary system, which is the endocrine component of the digestive system.

Exocrine and endocrine functions coexist in the pancreas. Endocrine organs secrete hormones directly into the bloodstream, while exocrine organs secrete hormones directly into the lumen (cavity) of another organ. The endocrine pancreas secretes hormones into the bloodstream. The exocrine pancreas comprises 99% of pancreatic tissue. It secretes digestive juices into the duct system and into the duodenum. Secretory cells are arranged in acini around small ducts and secrete pancreatic juice into progressively larger ducts.

The major innervation of the pancreas is parasympathetic, through the vagus nerve (cranial nerve X), and promotes secretion of digestive juices. Sympathetic stimulation inhibits secretion of digestive juices.

**The liver**

The liver is the heaviest and largest gland in the body (3 pounds). It is the second-largest organ after the skin. Its multiple functions include digestion, *hematopoiesis* (production of blood cells), detoxification, immune response, and synthesis of many chemical substances.

The liver occupies almost the entire right upper quadrant of the body. It is bounded by the diaphragm, above which it pushes up into the chest. Its edge is just at the right costal (rib) margin, and it is divided into a large right and small left lobe. The minor lobes are the *quadrate* and *caudate*, and the falciform ligament divides the right and left lobes. The liver is suspended from the back of the diaphragm by both the falciform and the suspensory ligaments.

The right branch of the hepatic artery (a branch of the celiac trunk) serves the right lobe, and the left branch serves the left lobe. There are very frequent (and surgically dangerous) variations in the location and size of the arteries. Systemic venous drainage occurs through the inferior vena cava, which traverses the liver and receives venous drainage from the hepatic veins. There are no other systemic veins because of portal circulation.

All the physiology and pathology of the liver depends on hepatic portal circulation. The portal vein drains the inferior mesenteric vein, the superior mesenteric vein, the splenic vein, the gastric veins, and the esophageal veins. Therefore, all veins draining the organs of digestion pass blood through the portal system in the liver.

The portal system divides into a capillary bed in the liver sinusoids. It re-forms on other side of the sinusoids as the hepatic capillaries and veins, which drain into the vena cava. It is a venous-capillary-to-venous-capillary system.

The microscopic structures of the liver are as follows:

- *Hepatocytes* are the liver cells that do all the work of the liver physiology.

- *Sinusoids* are small swellings between hepatocytes that act like small capillaries.

- The *portal triad* is the functioning unit. It consists of the hepatic artery, the portal (central) vein, and the bile duct. The routing is as follows:

    - Nutrient-rich blood enters the liver via the portal vein.

    - Blood reaches the hepatocytes by detouring through capillaries at the sinusoids, where exchanges take place.

    - Blood returns to the central (portal) veins.

    - Blood reenters circulation via the hepatic veins and the inferior vena cava.

    - Blood returning to the heart from the lower body through the systemic circulation and vena cava is not altered in transit.

All the hepatic (liver) cells are intimately intertwined with blood vessels. This has important implications for surgery and other treatment. The liver also produces bile, a digestive juice that emulsifies fat and excretes it into the hepatic ducts. Some bile is stored in the gallbladder for later use.

**The gallbladder**

The gallbladder is a secondary outpouching that develops from the liver. It lies in a groove under the liver, between the two lobes. It is a soft, thin-walled sac, shaped like a fat carrot, with its narrow end pointing toward the bile ducts (midline). Blood is supplied to the gallbladder by the cystic artery, which branches from the hepatic artery. It has a variable (surgically dangerous) position and origin.

Bile drains from the bile ducts of the liver into progressively larger ducts.

- The right and left hepatic ducts join just outside the liver to form the common hepatic duct.
- Bile from the gallbladder exits and enters through the cystic duct.
- The cystic duct and common hepatic duct join to form the common bile duct.
- The common bile duct goes behind the duodenum to join the pancreatic duct and enter the duodenum through the ampulla of Vater.
- The sphincter of Oddi, within the ampulla of Vater, can close off the flow of bile, which will then reflux into the gallbladder for storage.

The gallbladder is a vestigial organ and, as such, does not empty completely. Bile has many solids dissolved in it, which can precipitate out of solution, causing gallstones, infection, inflammation, and even cancer.

**Pancreatic cancer**
The ampulla of Vater can be involved in pancreatic cancer, which has a high lethality rate. Pancreatic tumors can block the bile duct and cause bile reflux into the blood and tissue (jaundice).

Cancers of the tail and body of the pancreas produce no symptoms until they are far advanced. Surgical treatment of pancreatic cancer involves removing the pancreas, duodenum, the bile ducts, and half the stomach and reconnecting the remaining organs (the Whipple procedure). Treatment of pancreatic cancer is especially difficult because the retroperitoneal location means that tumors tend to spread to highly innervated regions of the back and spine. ■

## Questions to Consider

1. Describe the structure of the portal circulation of the liver. Why is this complicated anatomy a necessary evolution of the digestive system?

2. Trace the route of venous blood in the colon as it enters the inferior mesenteric vein until it comes into the right heart.

# Digestive System—
# Physiology of the Pancreas, Liver, and the Biliary Tree
## Lecture 17

**Today we're going to continue with the physiology of the pancreas, the liver, and the biliary tree and see how the processes that we described anatomically come to play in what will be mostly the molecular biology and the chemistry of digestion.**

The pancreas and liver secrete digestive juices and enzymes that aid in digestion and absorption. In this lecture, we review the components of the exocrine pancreas, the main pancreatic hormones, and the mechanisms by which pancreatic enzymes are secreted into the small intestine. After reviewing several pancreatic disorders (pancreatitis and pancreatic cancer), we turn to the liver and examine the various functions that it performs. We conclude with a review of common liver disorders, notably jaundice, trauma, hepatocellular carcinoma, cirrhosis, and cholelithiasis.

### Primary processes of digestion and absorption

The primary processes of digestion and absorption occur in the small intestine. However, these functions depend heavily on the digestive juices and hormones secreted by the pancreas and the liver. The exocrine pancreas has the following components and functions:

- The pancreas produces 1,000–1,500 ml of pancreatic juice. per day.
- It consists mainly of water, NaCl (sodium chloride, or salt), and $NaHCO_3$ (sodium bicarbonate).
- Sodium bicarbonate buffers the juice and makes it alkaline (pH 7.1–8.2).
- The alkalinity stops the action of gastric pepsin and prepares chyme for the milieu of the small intestine.

**Pancreatic hormones**

Amylase digests the remaining carbohydrates into simple sugars. It is released into the blood in cases of pancreatitis. Amylase acts on three main sugars. These sugars are composed of other, simpler sugar molecules.

- Maltose is acted on by maltase and is broken down into two molecules of glucose.
- Sucrose is acted on by sucrase and broken down into glucose and fructose.
- Lactose is acted on by lactase and broken down into glucose and galactose.
- Trypsin, chymotrypsin, and elastase all digest proteins.
- Lipase digests triglycerides into fatty acids and monoglycerides.

**Physiological considerations**

Almost all pancreatic enzymes are secreted in an inactive form to prevent autodigestion (*pancreas* literally means "eats all flesh"). Inactive forms end in *–gen*, e.g., trypsinogen. In severe pancreatitis, activated enzymes may travel back into the pancreas and digest it. Acinar cells also contain a trypsin inhibitor that inactivates any active trypsin accidentally released in the pancreatic tissues. Enterokinase (from the small intestine) activates the enzymes once they are in the safe confines of the small intestine.

**Basic pancreatic function (chemical production)**

Food consists of generally complex molecules that are broken down into simple molecules and reassembled into necessary compounds by the pancreas, liver, and other organs. The body has evolved to favor molecules with similar structures. For example, the cyclopentanophenanthrene ring structure is common to cholesterol, estradiol, testosterone, and cortisol. The only differences between these compounds are one or two groups attached to the outside of a ring structure. However, these compounds have enormously different functions in the body. The body can replace a particular missing compound by modifying the creation process of a similar compound.

**Regulation of pancreatic secretion**

Parasympathetic stimulation to the pancreas (via the vagus nerve) occurs in response to the digestive processes of the stomach. It stimulates the secretion of all pancreatic enzymes. This is a preparation response.

The acid chyme enters the duodenum, along with partially digested fats, proteins, and carbohydrates. Enteroendocrine cells of the small intestine release cholecystokinin (CCK) and secretin. These two enzymes circulate into the bloodstream. These enzymes stimulate secretion of pancreatic enzymes into the small intestine and raise the pH. Secretin decreases gastric secretion; CCK inhibits gastric emptying.

**Cancer of the pancreatitis**

The causes of the rising incidence of pancreatic cancer are not known, although this disorder is linked to smoking. There are some 25,000 cases per year, with 95% mortality.

**The liver's multiple functions**

The portal circulation is the place of absorption and excretion of the liver. The liver has a reserve capacity of some 50–80%. The liver is one of the few human organs that can regenerate itself. Though it grows faster than any known cancer, regeneration does not become malignant, and the liver will stop growth at approximately its normal size. It is a vital organ; without it, the organism will die within days.

The liver synthesizes proteins:

- It makes almost all prothrombin and fibrinogen (clotting factors), as well as albumin, the major blood protein.
- It synthesizes proteins from amino acids.
- It also converts forms of amino acids when needed for specific proteins.
- It converts toxic ammonia (from amino acid conversions) into less toxic urea (which is excreted).

- It uses amino acids and proteins for energy production or storage as fats and carbohydrates.

The liver is the storehouse of carbohydrates as glycogen (glycogenesis) and lipids (lipogenesis).

- It can rapidly break down large amounts of carbohydrate (glycogenolysis) and release it as glucose into the bloodstream.

- It can create glucose from lactic acid (gluconeogenesis).

- The liver can store fats in various forms.

- It can break down and release stored fat for extraordinary needs.

- It synthesizes cholesterol from fatty acids and vice versa.

- Current therapy for high cholesterol uses drugs to block hepatocyte mechanisms for cholesterol synthesis.

Bilirubin removes broken-down pigments from dead red blood cells.

- It is metabolized with bile salts and excreted in the feces.

- In obstructive jaundice, bilirubin is not excreted, producing clay-colored feces.

The liver processes hormones and drugs.

- It detoxifies drugs and alcohol.

- Detoxification can be up- or downregulated to meet demand.

- The liver excretes processed toxins into bile and thence into the intestines for excretion.

- It alters the molecular structures of hormones to deactivate them.

The liver acts as a storage depot.

- It stores all the fat-soluble vitamins (A, B12, D, E, K).
- It stores iron and copper.
- The liver stores fat under stress or damage.
- Too much storage of fat, iron, or copper can damage liver cells.
- The bile salts emulsify fats to make them more water soluble and easier to absorb.

**Intestines**

The enteroendocrine glands secrete gastrin in response to gastric distension and the presence of proteins in the stomach. Gastrin stimulates the parietal cells ($H^+$ secreters) of the fundus of the stomach itself, increases gastric motility, and closes the lower esophageal sphincter.

**Small intestine**

Gastric inhibitory peptide is released from the duodenum and upper small intestine. Its target organ is the stomach, and is released by fatty acids and by incompletely broken-down carbohydrate. This hormone inhibits gastric secretion, slows stomach emptying, allows the small intestine to get ready, and stimulates insulin release from the pancreas (endocrine) to prepare for absorption of glucose.

Secretin is released from the small intestine. Its target organs are the pancreas, liver, and gallbladder. It is released by acid chime. It increases the bicarbonate in pancreatic juice and thereby increases the pH to neutralize acid. It also inhibits gastric juice secretion and maintains the pancreas.

CCK is also released from the small intestine. Its target organs are the pancreas, liver, and gallbladder. It is released by fatty acids in the duodenum and by partially broken-down proteins. It increases bile secretion from the liver, stimulates emptying of the gallbladder, closes the pylorus, and causes a feeling of satiety.

**Other minor hormones in the intestines**
Vasoactive intestinal peptide turns off gastric acid secretion. Stimulators of intestinal motility include motilin, bombesin, and substance P.

**Things that can go wrong with the intestine**

- Trauma to the internal organs and intestines is often fatal because of the large blood supply to these systems.

- Hepatocellular carcinoma is rare in the United States but common in Africa and Asia. It is related to a combination of insults to hepatocytes:

    - Aflatoxin from fungus (*Aspergillus flavus*).

    - Hepatitis B virus infection.

- Cirrhosis of the liver is severe scarring of the liver parenchyma, leading to blood backflow. It is caused by alcohol, hepatitis, and parasites. It is untreatable and ends in liver failure and portal hypertension. Symptoms include:

    - Clotting defects.

    - Esophageal bleeding.

    - Jaundice.

    - Hepatic coma.

    - Death.

**Gallstones (cholelithiasis)**
Gallstones are related to cholesterol metabolic defects, obesity, and pregnancy. There are some 50,000 cases per year in the United States. The usual treatment is laparoscopic surgery to remove the gallbladder. ■

## Questions to Consider

1. Name three hormones secreted by the enteroenteric glands of the small intestine and describe their target organs and effects.

2. What are the major causes of obstructive jaundice? Which one would you choose to have, if you had a choice? Which would be your last choice?

# Digestive System— Anatomy of the Small Intestine, Colon, and Rectum

## Lecture 18

**Today we're going to continue our way through the gastrointestinal tract, this time beginning at the pylorus, which is the guardian of the stomach, and work our work our way distally, towards the far end, through the small intestine, colon, and finally end up at the anus.**

We turn now to the anatomy of the organs in which nutrients and water are extracted from the chyme for use by the body and by which the resulting waste material is excreted from the body. The small intestine (also known as the small bowel) is the organ in which most of the absorption of nutrients and water occurs. We review its anatomical divisions (the duodenum, jejunum, and ileum), blood supply, and microstructure, especially the cells of the microscopic villi where absorption occurs. We turn next to the large intestine, or colon, which absorbs remaining water and transfers the feces to the rectum for excretion. The lecture concludes by examining the complicated anal sphincter muscle that controls passage through the anus.

**Small intestine**

The small intestine is responsible for the major absorption of nutrients and water. It is approximately 24 feet long, though the length depends on how and when it is measured (i.e., ante- or postmortem). It begins at the pylorus and ends at the ileocecal valve.

The first part is the *duodenum* ("twelve" inches). The duodenum runs from the pylorus to the ligament of Treitz, where it pierces the peritoneum. It is a retroperitoneal organ.

The second part is the *jejunum* ("empty"). The jejunum runs from the ligament of Treitz to the mid–small bowel. It has numerous muscular folds called *plicae circulares*. It is entirely an intraperitoneal organ.

The third division is the *ileum* ("twisted"). The ileum runs from the mid–small bowel to the ileocecal valve at the colon. It is an intraperitoneal organ. Plicae circulares also prominent in the upper to mid-ileum. They increase surface area and cause a corkscrew motion of chyme.

**Blood supply of the intestine**

The duodenum is supplied by the gastroduodenal artery and by branches of the superior mesenteric artery. The jejunum is supplied by jejunal branches of the superior mesenteric artery. The ileum is supplied by the ileal, right colic, ileocolic, and appendiceal branches of the superior mesenteric artery. Superior mesenteric syndrome is a rare disease in which unfortunate placement of the superior mesenteric artery causes it to block the duodenum under heavy food loads.

**Microstructure of the intestine**

The small intestine has the same four layers as the rest of the GI tract, but they are modified for maximal absorptive power.

The villi (singular: *villus*, "tuft of hair") are fingerlike projections of the mucosa.

- The single villus is the functional unit for absorption.
- There are 40 villi/mm$^2$.
- Each villus contains an arterial and venous capillary and a lacteal (lymphatic equivalent of a capillary). The lymphatic system is a circulatory system that moves between cells, drains into veins in the neck, and can also absorb fat.

Absorptive cells have microvilli on their surfaces for increased surface area. These microvilli form the brush border (200 million microvilli/mm$^2$.). They are different from the microvilli on respiratory cells, which are used for moving foreign material out of the system.

Intestinal glands are located in the crypts of Lieberkuhn. The cells here secrete intestinal juices.

- Paneth cells are in the deepest part of the glands. They secrete lysozyme (a bacteriocidal enzyme), and they are phagocytes.

- Enteroendocrine glands are the deepest part of the glands. The cells here secrete three hormones: secretin (S-cells), CCK (CCK-cells), and gastric inhibitory peptide (K-cells).

- Brunner's glands are in the deepest part of the duodenal mucosa. They secrete alkaline mucous to neutralize acid.

- Goblet cells secrete mucous.

Peyer's patches are sections of lymphatic tissue that detect foreign elements in the GI tract and signal the immune system.

The muscularis mucosa has several layers. The outer layer is longitudinal and thin. The inner layer is circular and thick.

Paralytic ileus is the absence of normal GI tract muscle contractions (peristalsis) and can be caused by anything that irritates the peritoneum sufficiently.

**Colon, rectum, and anus**

The large intestine or colon is designed only for the absorption of water. It is 5 feet long by 2.5 inches wide (with extreme variation). It extends from the cecum to the rectum in the pelvis. It comprises several distinct sections:

- The cecum receives chyme from the terminal ileum. It is guarded by a fold of mucosa called the ileocecal valve, preventing fecal backflow into the small bowel.

- The vermiform appendix is attached to the tip of the cecum.

- The ascending (right) colon extends from the cecum to the hepatic flexure and may or may not be intraperitoneal.

- The transverse colon extends from the hepatic flexure to the splenic flexure. It is always intraperitoneal.

- The descending (left) colon extends from the splenic flexure to the sigmoid colon.

- The sigmoid ("S-shaped") colon follows the descending colon.

The lines of haustration indicate places where the circular muscles have been pulled tight. Longitudinal muscles contract in the colon to form three bands, called the *tinea coli*, which can indicate obstructions in the colon. The right hepatic flexure and the left splenic flexure contact the liver and spleen.

The vermiform and retrocecal appendices are vestigial organs that may become inflamed or infected and require surgery.

Malrotation of the gut (small intestine) places the cecum and appendix in abnormal abdominal spaces.

**Blood supply of the colon**

- The right colon is supplied by the right colic arteries of the superior mesenteric artery (SMA).

- The transverse colon is supplied by the middle colic artery of the SMA.

- The descending colon is supplied by the left colic branches and arcades of Drummond of the inferior mesenteric artery (IMA).

- The sigmoid colon is supplied by the sigmoid arteries of the IMA.

- The rectum is supplied by the superior branch and inferior rectal arteries of the IMA.

Extensive collateral circulation exists along the arcades of Drummond, making devascularization of the colon difficult.

- Aneurysms (dilations of blood vessels), often filled with clots and fat, can block the IMA and build up collateral circulation to compensate.

- Colon cancer spreads along the major blood vessels, but the collateral circulation allows their removal if needed.

**Microstructure of the colon**

The microstructure of the colon has significant differences from that of the small intestine. There are no villi. Glands penetrate into the surface, rather than project in the manner of villus fingers. There are no permanent plicae circulares.

The rectum begins as the descending colon exits the peritoneum. It is defined by its extraperitoneal-ness. It is guarded by the complex anal sphincter. This muscle prevents incontinence and is one of the most complex muscle sets in nature. It can selectively pass any combination of gas, liquid, or solid.

- The external sphincter is voluntary.

- The internal sphincter is involuntary.

- Tenesmus is the feeling of needing to defecate and can become chronic in cancer patients.

**Colon diseases and surgery**

Diverticulitis (infection of an outpouching caused by high pressure) and cancer are common colon problems.

There are several types of colon surgery.

- *–tome* means “to cut.”

- *–tomy* means “to remove by cutting out.”

- *–otomy* means "to cut open with the intent of closing the wound."
- *–ostomy* means "to cut open permanently or semipermanently."

In a colostomy, the transverse colon is attached to an opening in the skin for a time to bypass an affected part of the colon. Feces is excreted into an external bag.

Colitis (allergy to one's own colon) causes inflammation, is a cancer risk, and may require a permanent ileostomy. ■

## Questions to Consider

1. What are the divisions of the small intestine? Which are intraperitoneal?

2. What would be the effect of removing the entire colon and rectum? Would this be more or less difficult to live with than removing the small intestine? Why?

# Digestive System—Physiology of the Small Intestine, Colon, and Rectum

## Lecture 19

**We're going to conclude our look at the gastrointestinal tract now by reviewing some of the physiology of the small intestine and the colon, as well as the rectum.**

In this concluding lecture on the GI tract, we first examine mechanical and chemical digestion in the small intestine—the process by which the small-bowel segments crush and mix the chyme to facilitate absorption of proteins, lipids, carbohydrates, and water. We turn next to the large intestine or colon, examining the various reflexes that move feces into and through the colon for excretion through the rectum. Finally, we examine the physiology of defecation.

The greater omentum hangs from the bottom of the stomach and is composed mostly of fat. It stores fat and provides a rich blood supply to the stomach. The lesser omentum is an attachment of the peritoneum that lies between the liver and the upper edge of the stomach. It carries the vessels that run to the stomach and liver.

The entire small bowel (duodenum, jejunum, and ileum) is devoted to two processes: digestion and absorption. Digestion is divided into mechanical and chemical phases.

### Mechanical digestion

Circular muscles constrict and divide the small bowel into segments. A muscle then contracts between two other muscles and subdivides the segment. Relaxation allows the segments to coalesce. This is repeated many times per minute so that the chyme is moved back and forth in the same area. Localized contractions crush and mix food within that segment alone. This action mixes the chyme with intestinal juices and prolongs its contact with the absorptive surface of the small bowel.

Peristaltic contractions, coordinated by the myenteric plexus, move the chyme slowly along the length of the small bowel. Peristaltic activity is weak, which means that food stays in the small bowel for a relatively long time (4–6 hours). Peristalsis is completely under the control of the autonomic nervous system. Appendicitis and surgery can slow or stop peristalsis. Because "the solution to pollution is dilution," the body increases peristalsis and pours fluid into the intestines to eject and weaken toxins in cases such as cholera. Victims then die of dehydration from massive diarrhea.

**Chemical digestion**

Chyme in the small bowel is a conglomerate of partially digested carbohydrates, lipids, and proteins. Digestion must be completed in the small bowel, because the colon will not absorb nutrients to any significant extent. Crohn's disease can require the surgical removal of segments of intestine, sometimes leaving the patient with inadequate bowel to absorb nutrients, or short bowel syndrome. Specific chemical digestions include the following:

- Proteins
    - Polypeptides (short-chain amino acids) arrive from protein breakdown by pepsin in the stomach.
    - Breakdown continues in the small bowel by activated pancreatic enzymes: trypsin, chymotrypsin, and elastase.
    - These three are all necessary because they each act at different places in the amino acid sequences.
    - Brush border cells of the small bowel excrete more peptidases—aminopeptidase and dipeptidase—that complete the splitting of the amino acids.
- Lipids (fats)
    - Triglycerides (three fatty acids bound to glycerine) are split by pancreatic lipase in the small bowel.

  - Pancreatic lipase splits two of the lipids off the glycerine. The remaining one is a monoglyceride.

  - Some lingual and gastric lipases have already been at work, but the major job is in the small bowel.

  - Lipids are not water-soluble (that is, they do not dissolve in water).

  - Bile salts (from the liver and gallbladder) emulsify (break into small droplets) the fat for easier entry into water solution.

  - Gallstones can cause fatty stools when they block bile from breaking up fat.

  - Patients with *fistulas* (bile leaks) may have to drink bile or have it injected in order to digest properly.

  - Malabsorption prevents the body from receiving nutrients dissolved in fat.

- Carbohydrates

  - Carbohydrates are cleaved into sugars by pancreatic amylase. (Pancreatic lipase and amylase in the blood are used to measure abnormal function of damaged pancreatic cells.)

  - Alpha dextrinase secreted by the brush border cells acts on remaining carbohydrates, cutting off one sugar at a time:

    - Maltose is glucose and glucose, split by maltase.

    - Sucrose is glucose and fructose, split by sucrase.

    - Lactose is glucose and galactose, split by lactase.

  - If lactase is insufficient, lactose intolerance develops. Bacteria ferment the high levels of lactose, and excess gas is produced.

**Absorption**

About 90–95% of nutrition is absorbed in the small bowel. Absorption of small fragments occurs as follows:

- Carbohydrates
    - Monosaccharides (glucose, fructose, galactose) are actively transported with sodium.
    - This active transport requires energy but gives a net gain of energy.
- *Proteins*
    - These include single amino acids, dipeptides, and tripeptides.
    - They are actively transported across the duodenum and jejunum. Varying amounts of energy are required.
- *Lipids*
    - These include fatty acids, monoglycerides, and glycerol.
    - They are absorbed by diffusion of small particles across the brush border and by active transport.
    - Fats get into lacteals, too, and they travel through the lymphatics to the venous system.

**Water**

Water is reabsorbed by osmosis. The following is what is called a "water balance sheet."

Table 1. Production and intake

| | |
|---|---|
| Saliva | 1.0 liter |
| Swallowed liquids | 2.3 liters |
| Gastric juice | 2.0 liters |
| Bile | 1.0 liters |
| Pancreatic juice | 2.0 liters |
| Intestinal juice | 1.0 liter |
| Total | 9.3 liters |

Table 2. Recycled and excreted

| | |
|---|---|
| Small intestine reabsorption | 8.3 liters |
| Colon reabsorption | 1.0 liters |
| Excreted in feces | 0.1 liter |
| Total | 9.4 liters |

Even small imbalances between fluid intake and output can cause major problems. Diarrhea is a common symptom of disease and can kill patients through dehydration. Rapid overconsumption of water or other liquids, although rare, causes a rapid drop in sodium and other electrolyte levels and can cause death.

**Regulation of secretion and motility**

The autonomic nervous system responds to the presence of chyme (distension). Note: Only distension causes pain in the small intestine.

- Myenteric plexus of the nerves signals states of distension.
- Parasympathetic impulses increase motility and secretion.
- Sympathetic impulses decrease motility and secretion.
- Vasoactive peptide (VIP) stimulates secretion.

**Absorption and storage**

The main jobs of the colon (large intestine), rectum, and anus are absorption of water and storage and convenient elimination of waste.

- When the stomach is distended, the gastroileal reflex signals the ileocecal valve to open and admit small bowel chyme into the colon.
- Haustral churning moves chyme around the colon slowly. This churning speeds up after meals, presumably to make room for a new load.
- During the filling of a segment, the haustra are relaxed.
- When a segment is distended, contraction occurs.
- Gastrin released in the stomach at mealtime also opens the ileocecal valve.
- The gastrocolic reflex moves chyme (which becomes semisolid and is called feces after 5–10 hours) quickly through the colon (*mass peristalsis*).

Final digestion is mediated by bacteria. Mucous is secreted to lubricate the colon, but enzymes are not secreted.

- Fermentation by bacteria produces methane gas, hydrogen gas, and breakdown of bile salts.
- Methane and hydrogen are flammable, which must be considered when using electrocoagulation in surgery.
- Gas is primarily swallowed air. Only 20% is produced from fermentation by bacteria.
- Distension of the cecum closes the ileocecal valve.

The feces is composed of water, salts, desquamated epithelial cells, bacterial decay products, and undigested food (fiber and so on).

- Feces remains in the colon for a long time, and carcinogens in feces (which are maximally concentrated at that point) may explain the prevalence of colon cancer.

- Bulky, indigestible fiber acts like a "colonic broom" to move feces through the system more quickly, carrying fat, cholesterol, and carcinogens with it.

- The feces end up in the rectum via mass peristalsis. The rectum is an excellent absorber: It can be used to instill water, salts, and medications rapidly.

- Receptors signal distension to the brain (conscious perception). Tenesmus is the feeling that one needs to defecate but cannot. Rectal tumors mimic that feeling.

- The defecation reflex is initiated when parasympathetic stimulation from the spinal reflex contracts the longitudinal rectal muscles.

    - Pressure increases in the rectum.

    - Pressure is added to the rectum by voluntary contraction of abdominal muscles.

    - Parasympathetic stimulation relaxes the internal sphincter of the anus (involuntary).

    - The external sphincter is opened by voluntary relaxation. This can be postponed by voluntary contraction.

    - Sphincter muscles weakened by age, disease, or trauma can cause *incontinence* (inability to hold feces in). ■

## Questions to Consider

1. Compare and contrast absorption in the small intestine with absorption in the colon.

2. What are the voluntary and involuntary components of defecation? What would be the effect of completely cutting the external anal sphincter?

# Endocrine System—The Pituitary and Adrenal Glands
## Lecture 20

**Today we're going to begin the first of our lectures on the endocrine system with the study of the pituitary and the adrenal glands. Now in these lectures we're going to study anatomy and physiology together each time so that we don't get too far away from the many, many glands that we're going to look at.**

This lecture is the first of three lectures on the endocrine system of glands that secrete hormones directly into the spaces surrounding their cells, from which the bloodstream picks up and circulates them. After distinguishing endocrine from exocrine functions and reviewing the endocrine system organs, we examine the functional differences between the endocrine and nervous systems and the basic properties of hormones. Next, we look at the most important endocrine glands: the pituitary gland and hypothalamus and the adrenal glands.

### Introduction and general definitions of the endocrine system

- Exocrine glands (*ex*, "out," *krinein*, "to secrete") are glands that secrete into ducts, which in turn, carry the secretions *out* of the glands and into the lumens of certain body cavities.

- Endocrine glands (*endo*, "within") are glands that secrete directly *into* the spaces around the cells and whose products are picked up and circulated by the bloodstream.

- The endocrine system includes some organs that are wholly endocrine in function. These include the pituitary gland, thyroid gland, parathyroid gland, adrenal glands, and pineal gland.

- Endocrine organs that have other functions as well as endocrine functions include the pancreas, liver, ovaries, stomach, hypothalamus of the brain, small intestine, kidneys, testes, and placenta.

There are differences between the functions controlled by the nervous system and the endocrine system. Both systems coordinate functions of body systems in general. Both are mutually interconnected.

**Endocrine system**

The endocrine system releases chemical messengers called *hormones* (*hormon*, "urge on"), which act on other organs in all parts of the body. Some hormones promote or inhibit nerve impulses, while others (epinephrine and norepinephrine) may be neurotransmitters themselves. These hormones act as hormones (rather than as neurotransmitters) in other places. Hormones may take seconds, minutes, or hours to work their effects, and their duration of action may be short- or long-lived. Hormones generally regulate growth, reproduction, and metabolism.

**Nervous system**

The nervous system acts by the generation of nerve impulses to stimulate or inhibit effector organs. It may stimulate the release or inhibition of hormones themselves from the endocrine organs. Nerve impulses have their effect in milliseconds, in contrast to the slower endocrine system, but the effects are also short-lived. Nerve impulses primarily cause muscle contraction and the secretion of fluids by certain glands.

**Endocrine gland locations**

The hypothalamus, pituitary gland, and pineal gland are in the brain. The thyroid gland is in the neck, with four parathyroid glands behind it. The thymus is in the chest. The adrenal (suprarenal) glands lie atop the kidneys. The stomach, ovaries, and testes in the abdominal cavity and lower also have endocrine functions.

**General properties of hormones**

Each of the 50+ hormones affect only a few cells, though they may reach all the cells of the body via the bloodstream. What accounts for the selectivity? Target cells contain highly specific receptors, which are surface glycoproteins. The geometry of the molecules allows only for very specific hormones to attach to the receptor in the target cell surface.

- In downregulation, each target cell has up to 100,000 receptors for a certain hormone. When there is an excess of hormone, the number of receptors decreases, reducing sensitivity.

- In upregulation, if a low number of hormone molecules is circulating, the number of receptors increases, raising the level of sensitivity.

Locally acting hormones do not enter the general circulation.

- Paracrine hormones (*para*, "near") act on cells next to the secreting cells without entering the bloodstream.

- Autocrine hormones (*auto*, "self") act on the cell that secreted them. Cancer cells use autocrine signaling to trigger growth.

**Pituitary gland and hypothalamus**

The pituitary is about 1 cm in diameter, and it lies in the *sella turcica* ("Turkish saddle") at the base of the brain, directly behind the optic chiasm. The pituitary gland, also called the *hypophysis*, was thought to be the "master gland" that controlled all the other endocrine glands. The hypothalamus actually controls the pituitary gland; it integrates many messages from parts of the brain and tells the pituitary what to do. Together, they regulate all processes having to do with primitive reactions, such as stress, rage, flight, body temperature, thirst, hunger, sexual activity, and survival in general.

A portal blood capillary system connects the hypothalamus and the pituitary. Portal flow allows blood-borne molecules from the hypothalamus to act on the pituitary before they are diluted with the blood in larger vessels. Between them, the hypothalamus and pituitary gland secrete 16 hormones.

The pituitary gland is divided into two embryologically and functionally different parts: the anterior pituitary and posterior pituitary. The anterior pituitary gland is also called the adenohypophysis. It makes up 75% of the pituitary gland. The anterior pituitary evolved anatomically up from the floor of the mouth (in contrast to the posterior pituitary [neurohypophysis], which evolved down from the base of the brain). Hormones released by the anterior pituitary flow into the general circulation for action in far parts of the body.

Seven releasing hormones are secreted by the hypothalamus and are responsible for the release or inhibition of the anterior pituitary hormones. These include growth-hormone–releasing hormone and growth-hormone–inhibiting hormone. They are generally controlled by negative feedback mechanisms. Anterior pituitary hormones are also controlled by negative feedback from the brain and the target organ.

The anterior pituitary hormones include:

- Thyroid-stimulating hormone (TSH), which stimulates the thyroid gland to release thyroid hormones.

- Follicle-stimulating hormone (FSH) and luteinizing hormone (LH), which together stimulate the release of estrogens and progesterones, which cause maturation of ova in the female and sperm cells and testosterone in the male.

- Prolactin (PL), which stimulates the production of milk by the breasts. It can cross the placenta-blood barrier, causing "witch's milk," or milk production from a baby's nipples.

- Adrenocorticotrophic hormone (ACTH), which stimulates the release of adrenal cortical hormones by the adrenal glands.

- Melanocyte-stimulating hormone (MSH), which causes increased skin pigmentation.

- Human growth hormone (hGH, or somatotropin), which stimulates body growth and regulates metabolic processes. High hGH increases the growth of the skeleton in the growing years of the child, and it maintains muscle and skeletal size in the adult.

The following abnormal conditions are associated with anterior pituitary hyper- or hyposecretion.

- *Pituitary dwarfism*: Low levels of hGH during the growth years causes bone-growth-plate closure before normal size is achieved. Many organs are small, and the person has a childlike stature. Synthetic hGH produced by recombinant DNA technology in bacteria has resulted in safe, plentiful sources and can prevent this if diagnosed in time.

- *Pituitary giantism*: Hypersecretion of hGH during childhood causes long bones and tall stature but otherwise normal proportions.

- *Acromegaly*: Usually caused by functioning pituitary tumors in the already normal adult; causes thickening of bones of the face, hands, and feet (bones can't get longer after closure of growth centers) and thickening of the tongue, eyelids, and nose. Goliath might have been an acromegalic giant. A pituitary tumor could place pressure on peripheral vision nerve fibers, causing tunnel vision, and a rock hurled from the side could hit the temple at the thinnest part of the skull.

The posterior pituitary gland (neurohypophysis) is anatomically derived from a downgrowth of the brain. It does *not* synthesize hormones, but it stores and secretes two of them. Hormones made in the brain are transported in small packets for storage in the posterior pituitary.

Oxytocin (*oxytocia*, "rapid child birth"; also, Pitocin) enhances the strength of uterine contraction and stimulates the ejection of milk after delivery. It may also foster maternal instincts and sexual pleasure during and after intercourse.

Antidiuretic hormone (ADH; also vasopressin) decreases urine production by increasing reabsorption by the kidneys. The effect is to raise blood volume and therefore blood pressure. Alcohol inhibits ADH secretion, thus producing profuse urination after a drinking binge and the headache and thirst associated with a hangover.

**Adrenal glands**

The adrenal glands are located almost directly on top of each kidney; hence, the terms *ad renal* and *supra renal*. They are small, retroperitoneal glands, about 5 cm in length, and weigh about 5 g each. They are supplied abundantly by three sets of vessels:

- Inferior phrenic arteries off the aorta.
- Middle suprarenal arteries off the aorta.
- Inferior suprarenal arteries of each renal artery.

Like the pituitary, they are composed of tissue from separate anatomic development during embryology.

- The adrenal cortex is the outer layer.
- The adrenal medulla is the inner layer.

The adrenal glands are absolutely essential for life. The adrenal cortex produces three hormones in three separate zones:

The first is mineralocorticoids. Aldosterone is 96% of this group, and it controls water and electrolyte (sodium and potassium) homeostasis. Their action is on the kidneys. Adrenal adenomas cause hyperproduction of aldosterone, which may account for 25% of hypertensive patients.

The second is glucocorticoids. Cortisol (also called hydrocortisone) is 95% of the total, plus corticosterone, cortisone. This drug depresses the immune system. It promotes protein catabolism (breakdown) and lipolysis—triglycerides to fatty acids. It promotes resistance to stress, resulting in higher blood pressure, and has anti-inflammatory effects. It retards allergic overreactions and slows wound repair. It promotes glucose formation (gluconeogenesis).

Addison's disease results from adrenocortical insufficiency. The results are lethargy, low blood pressure, weight loss, anorexia, and low blood sugar.

Addison's disease is treated with steroid hormone replacement. John F. Kennedy had Addison's disease and required regular cortisone injections to deal with stress. These injections changed his skin pigmentation, resulting in his constant deep "tan."

Cushing's syndrome results from excessive adrenal cortical function. It results in spindly arms and legs, moon-face, buffalo hump on the back, flushed skin, hypertension, osteoporosis, and decreased resistance to infection or stress.

Androgens are masculinizing hormones that occur in insignificant amounts in the adult male. In females, androgen accounts for sexual drive, and it is converted into female hormones (estrogens) after menopause. Old treatments for breast cancer involved removing the pituitary gland to prevent the adrenal glands from producing estrogen.

The third group of hormones are produced in the adrenal medulla. Hormones here are produced in the *chromaffin* cells ("color affinity"). They are innervated by the sympathetic division of the autonomic nervous system (ANS). Stimulation releases two hormones, called *epinephrine* (80%) and *norepinephrine* (20%), or *adrenalin* and *noradrenalin*. Collectively, they are called catecholamines. Thus, these cells are the postganglionic fibers of the sympathetic ANS.

Because the chromaffin cells are directly innervated by the preganglionic fibers of the ANS, they respond very quickly, as necessary in a system that responds to emergency situations. They continue to stimulate the secretion of adrenal hormones after nervous stimulation has passed. Unlike the adrenocortical hormones, medullary hormones are not essential for life in the quiescent state.

Hypersecretion by pheochromocytoma, a tumor of the adrenal medulla, causes paroxysmal hypertension, producing an extreme "fight or flight" reaction. ■

## Questions to Consider

1. How does the venous portal capillary system function in the anterior pituitary gland? Why is this necessary?

2. Describe the differences between endocrine control and nervous system control of bodily functions.

# Endocrine System—Pancreas

## Lecture 21

**Today we're going to continue our study of the endocrine system by looking at the endocrine function of the pancreas. It's a very large organ we've already looked at in its digestive form, or the exocrine function. Today we're going to look at it as purely an organ of endocrine secretion, secretion of hormones into the circulation.**

As an endocrine organ, the pancreas produces insulin and glucagon. After reviewing the four cell types composing the endocrine pancreas, the lecture examines in detail several insulin-related disorders: diabetes mellitus in its two principal types and hyperinsulinism. The pancreas functions in two modes: It is an exocrine digestive organ, and t is also an endocrine organ, producing insulin and glucagon.

The pancreas is nestled in the curve of the duodenum and stomach. It is entirely retroperitoneal. Its extensive blood supply comes through anastomoses of the gastroduodenal artery, the posterior superior pancreaticoduodenal artery, the anterior inferior pancreaticoduodenal artery, and the splenic artery. The head of the pancreas joins the second portion of the duodenum. The tail meets the spleen.

Ninety-nine percent of the pancreas is made of acini, clusters of exocrine digestive cells connecting the ducts. One percent of the pancreas is made up of several million scattered islets of Langerhans, cells that contain the endocrine functioning.

**Physiology of the endocrine pancreas—four cell types**

Alpha cells constitute 20% of the islet cells. They secrete the hormone glucagon, which raises blood sugar to maintain normal levels. Chemoreceptors measure the amount of sugar in the blood; low blood sugar directly stimulates the release of glucagon from the a-cells. Glucagon acts on hepatocytes to convert glycogen to glucose (glycogenolysis) and to convert amino acids into glucose (gluconeogenesis). Higher-than-normal blood sugar turns off the release of glucagon. Stimulation of the sympathetic nervous system in preparation for stress, fight, and flight also affects glucagon release.

Beta cells constitute approximately 80% of islet cells. The beta cells secrete insulin, which lowers blood sugar. Lower-than-normal blood glucose turns off the output of insulin. Higher-than-normal blood sugar stimulates beta cells to release insulin. High blood sugar is bad not only for the blood but also for cells, which do not receive the glucose they need. Genetically engineered bacteria now produce synthetic human insulin.

Insulin acts on body cells in the following ways:

- It increases the speed of diffusion of glucose into the cells (especially skeletal muscles).

- It accelerates the conversion of glucose into its storage form, glycogen.

- It increases synthesis of proteins from amino acids.

- It increases synthesis of fatty acids.

- It decreases the rate of glycogenolysis and gluconeogenesis.

Glucagon and insulin have a clinical application in the treatment of excess blood potassium:

- Potassium is an ion with high concentration in cells and low levels in the blood; sodium is an ion for which the opposite is true.

- Movement of these ions between cells and the blood enables electrical conduction down nerves and in the heart.

- Excess levels of potassium can fatally prevent electrical conduction.

- Simultaneous injection of glucose and insulin will drive excess glucose into cells, taking potassium with it and decreasing blood potassium levels.

**Other factors affect insulin release**

- Parasympathetic innervation to the islets stimulates insulin release (recovery and rest).

- Glucagon itself causes insulin release to balance its effect in a negative feedback loop.

- Gastric inhibitory peptide (GIP) from the enteroendocrine cells of the small bowel responds to glucose in the lumen of the gut.

- Certain amino acids also affect insulin release.

**Delta cells**

Delta cells constitute less than 1% of pancreatic islets. They secrete somatostatin (growth-hormone–inhibiting hormone; the same as that secreted by the hypothalamus). This hormone inhibits insulin release and slows absorption of nutrients from the GI tract.

**Insulin-like growth factors**

Insulin-like growth factors, found in many body tissues, share the molecular structure and shape of insulin and are involved in growth. They are used both clinically and by cancer cells to stimulate growth.

**F cells**

F cells constitute less than 1% of islet cells. They release pancreatic polypeptide, which inhibits the release of somatostatin.

***Diabetes mellitus* ("copious sweet urine")**

Diabetes mellitus is a group of disorders in which glucose levels are elevated in the blood. It was originally diagnosed by tasting the patient's urine; elevated glucose levels make the urine sweet. It is characterized by the three "polys":

- *Polyuria*: copious amounts of urine outflow.
- *Polydipsia*: excessive thirst and drinking (water).
- *Polyphagia*: excessive eating.

There are two main types of diabetes mellitus. Type I is insulin-dependent diabetes mellitus and represents about 10–20% of all diabetes cases. With this type, there is an absolute deficiency in the quantity of insulin available. It is also known as *juvenile-onset diabetes*, because it often presents in childhood. Patients are referred to as "brittle" diabetics, because their dietary and insulin demands fluctuate and are hard to meet. It may be an autoimmune disease in which the body becomes allergic to its own beta cells and destroys them. The patient is in a state of starvation, unable to use nutrients without injections of insulin. Cells use fatty acids to produce energy (because they cannot use glucose); the result is an excess of fatty acid wastes called *ketones*.

Ketones are acidic, and they cause a shift to acidity in the blood. This condition is called *ketoacidosis*. Uncorrected, ketoacidosis is rapidly fatal. Low- or no-carbohydrate diets cause the body to selectively burn fat by interrupting the Krebs cycle, which produces energy but needs a primer of 100 grams of glucose. This process essentially turn dieters into diabetics and produces a mild form of ketoacidosis, which can (and has) become fatal. If a dieter eats protein and fat, *then* triggers the Krebs cycle, all excess material will be turned into fat anyway.

Type I diabetes is called a protean disease because it affects every system in the body. Excessive fat in the blood over long periods leads to atherosclerosis. Other physical defects related to high blood lipids and blood vessel damage include strokes, heart attacks, kidney failure, peripheral vascular disease, and increased rates of infection. Both cataracts of the lens of the eye and diabetic retinopathy are related to high blood sugar. Standard treatment is daily insulin injections; experimental treatments include pancreatic transplants, islet cell transplants, and immune suppression.

Type II is non-insulin–dependent diabetes, also known as *maturity-onset diabetes*. This type accounts for 80–90% of all cases. Its victims tend to be more than 40 years old, and many are obese. Type II diabetes is much milder than type I diabetes and easier to control. Many patients have normal insulin levels, but target cells are less sensitive to insulin action. Symptoms include the three polys mentioned above. Treatment options include diet, weight loss, and exercise, which alone works for many patients; insulin-stimulating drugs, which are necessary for some; and insulin injections, which are necessary for a few. External insulin pumps are another option.

Gestational diabetes usually results from transient elevations in blood glucose during pregnancy. It causes the same problems as type II for the fetus.

**Clinical pictures of out-of-control diabetes**
Patients with high blood sugar/too little insulin exhibit slow onset, lethargy, confusion, odor of acetone on breath, coma, and death. Treatment is insulin. Patients with low blood sugar/too much insulin have a faster onset of symptoms, leading in some cases to confusion and coma, shock, and death. Treatment is oral or intravenous glucose. Patients with overdoses of narcotic drugs can present similar symptoms (lethargy, confusion, coma). Injection of Narcan will immediately reverse the effects of the narcotic but leads to wild behavior as the patient goes into instant withdrawal.

**Tumors of the endocrine system**
Benign tumors of the endocrine system can become autonomous and release large amounts of unwanted hormones into the body. Zollinger-Ellison syndrome is caused by a gastrin-secreting tumor in the pancreas, which causes uncontrollable, seemingly incurable ulcer disease from overproduction of stomach acid. Zollinger-Ellison was formerly treated by removing the stomach's parietal cells but can now be treated with proton-pump inhibitors that block acid production. ■

## Questions to Consider

1. Define and describe the differences between exocrine and endocrine glands. In what category is the pancreas?

2. List as many symptoms as possible in a diabetic patient who has forgotten to eat after taking a normal dose of insulin.

# Endocrine System—Thyroid and Parathyroid Glands

## Lecture 22

**You may have noticed a peculiarity with the endocrine organs: We tend to have a lot of roommates that have entirely different jobs. ... Now we're going to look at the thyroid and the parathyroid glands, which also have completely separate functioning but are intimately related in geography.**

We finish our survey of the endocrine system with the thyroid and parathyroid glands. The thyroid gland regulates the rate and intensity of the body's chemical reactions, and the parathyroid glands regulate the amount of calcium and phosphorous in the blood.

For each of these glands, we briefly review the gross and microscopic anatomy, the physiology of the gland, and the consequences of dysfunction.

Thyroid dysfunction (either hypo- or hyperthyroidism) can lead to cretinism, myxedema, Graves' disease, and other pathologies. Parathyroid dysfunction (either hypo- or hyperparathyroidism) can lead to various disorders, including overexcitement of the muscle and nervous systems, bony demineralization, high calcium levels, duodenal ulcers, kidney stones, and behavioral disorders.

### Thyroid gland

The thyroid gland is the thermostat of the body, regulating the rate and intensity of chemical (metabolic) reactions. It is an H- or butterfly-shaped gland located just below the larynx (voice box); the right and left lobes are connected by an isthmus. It envelops the trachea anteriorly (clinically significant in cases of thyroid cancer).

The gland is very richly vascularized via branches of the carotids and the thyrocervical trunk, including the superior thyroid artery, the inferior thyroid artery, and multiple thyroid veins. The thyroid has clinically important relationships with the recurrent and superior laryngeal nerves.

The thyroid gland is made primarily of spheres called *follicles*. Two types of cells compose the follicles: The follicular cells produce thyroxine (tetraiodothyronine or T4) and triodothyronine (T3). The parafollicular cells (C cells) produce calcitonin.

Thyroid chemistry is iodine-based; iodine must be ingested because it is an element. Benign goiter (enlargement of the thyroid) is common in areas where iodine does not naturally occur in food. Western countries typically include iodine in normal table salt. Cancerous thyroid tumors (nodules) normally occur in patients whose faces have been irradiated, but these tumors are easily curable.

**Thyroid hormone synthesis**

Hormones made in the cells are stored in the follicles in large quantities as colloid for later use; about a three-month supply is stored. At all levels of function, thyroid-stimulating hormone (TSH) from the anterior pituitary regulates the processes via a negative feedback loop. Follicular cells actively trap virtually all iodide molecules in the body.

Thyroglobulin is synthesized from amino acids and iodide and stored in the follicles as colloid. When needed, T3 and T4 are released into the circulation from the follicular cells. T4 is present in much greater quantity than T3, but T3 is more potent. T4 is converted to T3 in the body cells.

The thyroid's ability to trap iodine can be used clinically. Low levels of radioactive iodine (I-131), combined with X-ray exposure, can map thyroid function. Higher levels of I-131 will irradiate and destroy thyroid tissue, when needed, without damaging surrounding cells.

Cancerous cells will not trap iodine when normal thyroid tissue is present. Tumors must be surgically removed or thyroid tissue must be destroyed before I-131 can destroy cancerous thyroid cells.

**Thyroid hormone action**

Thyroid hormones regulate the following activities:

- Oxygen uptake.
- Basal metabolic rate (gross); maintenance of body temperature.
- Intracellular metabolism (microscopic) protein synthesis, lipid breakdown, and cholesterol breakdown.
- Growth and development; that is, body growth rate and nervous system development.

Thyroid hormone also enhances the effects of catecholamines, accounting for high blood pressure, nervousness, sweating, and fast heart rate in hyperthyroid patients.

**Iodine uptake and control**

Iodide ($I^-$) ions circulating in the blood are actively taken into follicular cells through capillaries and become trapped in the endoplasmic reticulum. The follicles begin synthesizing thyroglobulin. Vesicles (small transport membranes) transport iodide further into the follicles, where it is combined with thyroglobulin to produce the amino acid tyrosine.

Tyrosine is bound into colloid, where it can be transformed into needed compounds (T3, etc.). Negative feedback loops between the thyroid and pituitary gland control the output of TSH. High levels of iodine in the thyroid gland inhibit the release of TSH.

**Pathology and thyroid dysfunction**

- Cretinism: Hypothyroidism during fetal development leads to dwarfism, mental retardation, and physical deformities.
- Myxedema: Hypothyroidism during adult life leads to edema, slow heart rate, low blood pressure, lethargy, sensitivity to cold, low body temperature, muscle weakness, hair loss, and obesity but not mental

retardation. *Myxedema madness* refers to occasional psychotic episodes that this condition may produce.

- Hyperthyroidism causes increased heart rate, increased blood pressure, high body temperature and sweating, nervousness, diarrhea, heat intolerance, and weight loss despite high caloric intake.

- Graves' disease is an autoimmune disorder in which antibodies mimic the effects of TSH but are not constrained by the negative feedback system for turn-off and control. This disease causes goiter, enlargement of the thyroid, and exophthalmos (bulging eyeballs caused by deposition of fat behind the eye). Curing the disease may not cure exophthalmos, which may leave the eyes open to injury.

**Parathyroid gland**

The four *parathyroid* ("beside the thyroid") glands regulate the levels of calcium and phosphorous in the blood. They are located on both sides of and behind the thyroid and have unrelated functions but are intimately connected to the covering of the thyroid gland.

There are two parathyroid glands on each side of the thyroid, named the *right* and *left inferior* and *superior glands*. They are supplied by the thyroid vessels. Each is about the size of a small pea or a large kernel of rice. They can be difficult to identify.

The chief cells (*principal cells*) produce parathormone (PTH). The function of the oxyphil cells is unknown.

**Physiology of PTH**

PTH raises serum calcium and phosphorus levels in the blood. PTH increases the cells of the bone (osteoclasts), which reabsorb calcium. It increases urinary reabsorption of calcium by the kidney. It also increases excretion of phosphorus by the kidneys (which, in turn, increases calcium levels). It causes kidneys to form calcitrol, a hormone made from vitamin D that increases the absorption of calcium from the GI tract.

Calcitonin (from the thyroid gland), on the other hand, participates in the negative feedback system by forcing calcium back into the bones.

**Pathology of parathyroid dysfunction**

Hyperparathyroidism is increased PTH production, usually because of a benign tumor of one or more of the parathyroid glands (parathyroid adenoma). Calcium is reabsorbed from the kidneys, bones, and stomach into the blood. "Stones, bones, groans, and moans" refers to a classic set of four symptoms: kidney stones, demineralized bones (osteoporosis), groans of pain from duodenal ulcers, and "moans" of psychosis. Removing a parathyroid adenoma can cause an immediate and drastic return to normal function and the disappearance of all symptoms. In parathyroid hyperplasia, all four glands overproduce PTH. Removing most of the parathyroid glands can fix the problem, but errors can cause hypoparathyroidism.

Hypoparathyroidism leads to low serum calcium levels and an elevated state of excitement for nerves and muscles, resulting in twitching and overactivity of the muscular and nervous systems. In the extreme, this can lead to convulsions and death. It is caused primarily by inadvertent surgical removal. Removed parathyroid glands can be chopped up and implanted into muscle in other areas of the body, where sometimes they will survive and start producing PTH again. Hypoparathyroid patients otherwise need lifelong calcium and vitamin D injections and are hard to manage.

Parathyroid glands can migrate down into the chest cavity, making them difficult to find and remove. ■

## Questions to Consider

**1.** Describe the function of the thyroid gland.

**2.** What is a negative feedback mechanism? Describe one such system in detail.

**3.** What is a goiter? What is exophthalmos? Are they related? If so, how? If not, why not?

# Urinary System—
# Anatomy of the Kidneys, Ureters, and Bladder
## Lecture 23

**We're going to move on now to the anatomy of the kidneys, the ureters and the bladder which comprise the urinary excretory system. … The urinary system is primarily one of excretion as well as an equal part conservation and then convenience for evacuation of the excreted materials.**

This is the first of two lectures on the urinary system. In this lecture, we will examine the anatomy of the kidneys (which filter out wastes and recycle needed elements), the ureters (which convey wastes from the kidney to the bladder), and the bladder, which stores wastes until excretion. After reviewing the gross anatomy of the kidneys, we consider the kidneys' major functioning unit—the nephron, of which there are some 1 million in each kidney. Each nephron is composed of a renal corpuscle, a renal tubule, and collecting ducts. The lecture concludes by reviewing the conduits through which urine passes before excretion from the body.

The urinary system is composed of the kidneys, ureters, bladder, and urethra. The ureters are conduits to the bladder from the kidneys. The bladder stores urine and wastes until a convenient time for disposal. The urinary system also has functions not related to waste elimination and recycling. These functions include maintaining blood volume, normal blood pressure, normal blood composition, and normal body and blood pH, as well as synthesizing calcitrol and secreting erythropoietin.

### Kidneys

The kidneys are paired organs lying to the right and left of the spinal column. They are entirely retroperitoneal. They are each supplied by one blood vessel (an "end vessel"), with no collateral circulation. They are necessary organs. Malfunction, however, does not cause problems on such an acute (short time frame) basis as, for example, the brain; kidneys have a chronic (long time frame) rate of pathology. Patients with total kidney failure can survive for several days.

The kidneys' primary function is to maintain homeostasis. They filter and excrete wastes, solutes, and toxins and absorb water, electrolytes, and buffers. They also recycle necessary elements and maintain normal blood volume, blood pressure, and pH. They synthesize calcitrol and erythropoietin, which stimulates production of red blood cells.

Word beginnings referring to kidneys include *renal–* and *nephro–*. They are protected by the ribs and the thick back muscles. The simplicity of the kidneys' blood supply makes transplantation relatively easy. A frontal section of the kidney would reveal:

- The outer layer, called the *renal cortex*.
- The inner layer, called the *renal medulla*.
    - The latter consists of 10–15 medullary pyramids.
    - The medial tip of the renal pyramid is the *renal papilla*.
- The *renal pelvis* is composed of *calyces* (*calyx*, "cup"), minor calyces, and major calyces.
- The *renal hilus* is the entryway for the vessels and ureter. It faces the medial side of the kidney.

**Vascular supply**

- Renal arteries—usually one to each kidney, coming directly from the aorta.
- Renal veins—usually one each, draining directly into the inferior vena cava.
- The blood vessels branch out inside the kidneys and create internal collateral circulation. This makes it possible to remove sections of a diseased kidney.

### Kidneys—microscopic anatomy

The *nephron* is the major functioning unit of the kidney. There are approximately 1 million nephrons per kidney. One's complement is complete at birth, and the loss of nephrons can be made up without new ones being formed. Patients can survive after losing an entire kidney, because the other kidney can increase its workload.

### Nephron composition and function

The renal corpuscle is composed of an afferent arteriole, the *glomerulus* ("little ball"), and an efferent arteriole. The renal tubule and collecting ducts are composed of peritubular capillaries, the vasa recta, and the renal tubule. The microstructures of the renal tubule and collecting ducts include the proximal convoluted tubule (PCT), the loop of Henle (nephron loop), the distal convoluted tubule (DCT), the collecting ducts, and the juxtaglomerular apparatus. There is a unique blood flow from the arterial capillary to arterioles, unlike portal systems, where the flow is from the venous capillaries to venules. The afferent arteriole is larger than the efferent, creating a forward pressure gradient.

### Microscopic function

The glomerulus is a group of capillaries covered in a double layer (visceral and parietal) of epithelium, called the *Bowman's capsule*. Blood runs across the Bowman's capsule and into the proximal convoluted tubule. The proximal convoluted tubule is lined with blood vessels that reabsorb nutrients from it and excrete wastes into it.

- The juxtamedullary nephrons extend deep into the medulla and have much longer loops of Henle than normal nephrons to fine-tune blood composition.

- The endothelium of the glomerular capillaries is fenestrated (it has small windows or holes) to retain red and white blood cells and allow everything else to escape.

- The basement membrane has smaller windows that also trap larger proteins.

- Podocytes (“foot cells”) have even smaller holes that trap smaller proteins but allow ions to pass.

- Fluid entering the proximal convoluted tubule should then be free of blood cells, proteins, and other large molecules.

**Ureters**

The ureters are paired retroperitoneal organs. They exit the kidneys at the renal hilus and course downward, parallel to the spinal column. The ureters are easy to cut in surgery without realizing it, which can be fatal. The ureters can peristalse to move urine without gravity. Ureter distension (stretching) or stones can cause excruciating pain, on a level with or worse than labor in childbirth. However, stone pain is considered much worse psychologically, because it is unpredictable and has no positive emotional affiliations. The ureters enter the bladder at the trigone. They form a functional (not anatomical) valve to prevent urine reflux.

**Bladder**

The bladder is located in the midline behind the pubis. It is located in front of the rectum in the male and in front of the uterus in the female. It has an internal sphincter (involuntary) and an external sphincter (voluntary). The urethra exits from the bladder. It is short in the female and longer in the male penis. The shorter length of the female ureter makes incontinence and bladder infections more common in females.

**Dialysis**

An osmotic pump that mimics the glomerulus can filter a patient’s blood externally to replace damaged or destroyed kidneys. Dialysis must be performed every two to four days, but patients can lead otherwise normal lives. ■

## Questions to Consider

1. Describe the anatomy of a nephron in detail.

2. Describe the internal blood flow of the kidneys. How is it different from a venous portal system?

# Urinary System—
# Physiology of the Kidneys, Ureters, and Bladder
## Lecture 24



**We're just going to look at the microscopic physiology of the functioning unit of the kidney, which is the nephron. Now, we tend to think of this organ in general terms as an excretory organ. But it's really one that's involved primarily with homeostasis, maintenance of a stabile, internal milieu. And the excretion is really a byproduct of that maintenance of homeostasis.**

The primary function of the urinary system is to maintain the body's homeostasis. This lecture focuses on the physiology of the nephron, specifically on the process by which the nephron filters out many of the blood's components, reabsorbs some, and removes others. The stages involved in this process are glomerular filtration, tubular reabsorption, and tubular secretion. The lecture concludes by briefly reviewing the physiology of the ureters and bladder.

### Introduction to the kidneys

The primary function of the kidneys is to maintain homeostasis. General mechanisms for maintaining homeostasis include the following:

- Regulation of water volume, blood volume, and interstitial fluid volume.
- Reabsorption of required substances.
- Excretion of excess substances.
- Excretion of toxic substances.
- Regulation of body pH.
- Regulation of normal blood pressure.

Osmosis is an important process in the urinary system and the body in general. Osmosis is defined as the passage of water molecules across a semipermeable membrane from a lower to a higher concentration of water molecules. Water on the side of higher concentration produces pressure on the membrane. The membrane allows water to pass but not dissolved solids.

**The nephron**

The nephron is the largest functioning unit of the kidney. The glomerulus and its arteriole are contained in the capsule of Bowman. Filtrate passes through the glomerulus and into the proximal convoluted tubule. The proximal convoluted tubule passes filtrate downward through the loop of Henle, then back up. The juxtaglomerular apparatus in the loop of Henle is involved in the release of certain important hormones.

**Glomerular filtration**

- Pressure pushes fluid and solutes (the glomerular filtrate) through microscopic holes in the glomerular capillaries. Almost everything received by the nephron is passed into the glomerular filtrate.

- Tubular reabsorption in the convoluted tubule absorbs important solutes from the filtrate.

- In tubular secretion, wastes are pushed into the tubule for disposal.

- Normal blood pressure produces a pressure of about 55 mm Hg in the afferent arteriole; the glomerular filtrate exerts a capsular hydrostatic pressure of about 15 mm Hg in the opposite direction.

- Blood colloid osmotic pressure (BCOP), produced by the pressure of proteins and other solids in the blood, amounts to a pressure of 30 mm Hg against the arteriole pressure.

- The net filtration pressure—the force that pushes filtrate through the kidney—is 55 mm Hg – 15 mm Hg – 30 mm Hg = 10 mm Hg. This does not change much with normal blood pressure variation, but shock can reduce pressure and shut off arterial blood flow to the kidney.

The arterioles contain differently sized filters—the endothelium, the basement membrane, and the slit membrane—that filter out large proteins. Increased blood pressure increases the net filtration pressure to some extent. The two kidneys together receive 20% of the total blood flow. Almost 200 liters are filtered per day, but 99% is reabsorbed.

Net urine output is 1–2 liters per day. When the body is relaxed, the discrepancy in size between the afferent and efferent arterioles produces the net filtration pressure. Sympathetic stimulation constricts both vessels by the same amount, leaving the net pressure unchanged. Many of the blood's components are filtered: glucose, amino acids, small proteins, waste molecules (urea), and Na, K, Cl, and $HCO_3$ ions. Most of these substances are reabsorbed.

Glomerular filtration is regulated by three mechanisms with five control systems.

*Renal autoregulation* is the normal process at rest.

- Increased blood pressure stretches the walls of the afferent arteriole, causing smooth muscle in the arteriole walls to constrict.
- Constriction lowers the net pressure and the glomerular filtration rate (GFR).
- In tubuloglomerular feedback, high levels of sodium trigger the macula densa to constrict the arteriole.
- A decrease of nitric oxide will also constrict the arterioles.

In *neural regulation*, Increased sympathetic activity constricts the arterioles of the kidneys and releases norepinephrine and renin.

*Hormonal regulation* occurs through the action of several hormones:

- Angiotensin II (released by renin) constricts the efferent and systemic arterioles, increasing pressures throughout and decreasing GFR.

- Atrial natriuretic peptide (ANP) is produced when the left atrium of the heart is stretched. It circulates to the kidney and relaxes mesangial cells in the glomerulus, increasing absorptive capillary surface and GFR.

- Aldosterone is produced by the adrenal cortex in response to angiotensin II. It retains sodium and excretes potassium, causing osmosis that raises blood pressure.

Sustained or severe “fight or flight” reactions can stop blood flow to the kidneys for too long, killing the tubules (acute tubular necrosis, or ATN) and the entire kidney.

Urine output is a primary symptom of kidney pathology. Oliguria is defined as an output of less than 250 cc of urine per day. Anuria is defined as an output of less than 50 cc of urine per day.

**Tubular reabsorption**

Urine in the tubules must be reabsorbed to prevent dehydration, which would occur within hours. Solutes (glucose, amino acids, sodium, chloride, potassium, phosphate, and bicarbonate) are both actively and passively reabsorbed in the tubules.

*Absorption* is the taking in of substances from outside the body. *Reabsorption* involves transfer from one part of the body to another. Substances are filtered, reabsorbed, and excreted in urine:

**Table 3. Normal Reabsorption Rates in Humans**

| Substance | Filtered (enters glomerular capsule per day) | Reabsorbed (returned to blood per day) | Urine (excreted per day) |
|---|---|---|---|
| Water | 180 liters | 178–179 liters | 1–2 liters |
| Proteins | 2.0 g | 1.9 g | 0.1 g |
| Sodium ions ($Na^+$) | 579 g | 575 g | 4 g |
| Chloride ions ($Cl^-$) | 640 g | 633.7 g | 6.3 g |
| Bicarbonate ions ($HCO_3^-$) | 275 g | 275 g | 0.03 g |
| Glucose | 162 g | 162 g | 0 g |
| Urea | 54 g | 24 g | 30 g |
| Potassium ions ($K^+$) | 29.6 g | 29.6 g | 2.0 g |
| Uric acid | 8.5 g | 7.7 g | 0.8 g |
| Creatinine | 1.6 g | 0 g | 1.7 g |

Any excretion of glucose in urine is abnormal. Diabetes patients, for example, produce much more glucose than the tubules can handle; the excess is excreted in urine.

The descending loop of Henle reabsorbs water and large ions and molecules. Final adjustment takes place in the ascending loop. Aldosterone increases sodium and water reabsorption by the principal cells. Antidiuretic hormone (ADH; secreted by the posterior pituitary) facilitates uptake of molecules by the principal cells of the convoluted tubules.

**Water reabsorption**

Ninety percent is passive (obligatory) water reabsorption. It occurs by osmosis in the proximal convoluted tubule, the descending limb of the loop of Henle, and the distal convoluted tubule (DCT). Ten percent is facultative water reabsorption, aided by ADH and aldosterone. It occurs in the last part of the DCT and the collecting ducts.

**Tubular secretion**

Tubular secretion rids the system of unwanted materials. It occurs mostly in the tubules.

- Hydrogen ions ($H^+$) are removed when pH is too acidic. Tubular secretion acidifies urine by actively pumping out $H^+$ ions.

- Bicarbonate ions ($HCO_3^-$) are increased or decreased to maintain pH.

- Potassium ion ($K^+$) secretion is complex:

    - Aldosterone increases secretion of $K^+$ by tubules.

    - Increased $Na^+$ levels encourage secretion of $K^+$ in DCT.

    - High serum $K^+$ levels cause increases in tubular secretion of $K^+$.

    - This system is especially important because high levels of $K^+$ can rapidly stop the heart; potassium imbalances are the most dangerous problem for patients on dialysis.

**Clinical applications**

Before the advent of hemodialysis, fluid was pumped into the abdominal cavity for absorption through the peritoneum. However, this technique is prone to cause infection and becomes less effective over time.

A diuretic is anything that causes diuresis (outflow of large amounts of urine). Diuretics are commonly used to control blood pressure. For example, osmotic diuretics use direct osmotic pressure to create short-term diuresis; caffeine decreases sodium absorption, causing diuresis through osmosis; and Lasix (furosemide) works in the loop of Henle to produce massive diuresis. Lasix also increases blood flow to the kidney, which can be useful in its own right.

Urinalysis can detect many problems. Normal urine should be clear (not cloudy), concentrated, and contain no abnormal components, such as blood or protein. Tests measure urea, creatinine, nitrogen, and other components. Clearance tests compare the amount of waste products with the amount of those products in the blood and are very sensitive. ■

## Questions to Consider

1. Describe the mechanism of water reabsorption and name the locations for reabsorption in the nephron.

2. What is the effect of severe and prolonged hypotension (low blood pressure) on renal filtration? Renal blood flow?

# Reproductive System—Male

## Lecture 25

**Today we're going to begin the study of the reproductive system. We're going to focus today on entirely the male reproductive system and its function in the production, storage, and delivery of spermatozoa and the production, storage, and release of the male sex hormones.**

This is the first of three lectures on the reproductive system. This lecture examines the gross anatomy of the male reproductive system. The scrotum contains the testes, which produce spermatozoa through the process of spermatogenesis. The lecture also reviews the functions of the prostate and Cowper's glands, the process of erection and ejaculation, and the composition of the semen.

### Embryology of the external sex organs

Five-week-old embryos have undifferentiated sex organs: the genital tubercule, the glans, the urethral folds, the labioscrotal swelling, and the beginning of the perineum. At 10 weeks, some differentiation has taken place in the glans and the urethral folds, and the perineum has connected to the anus. Just before birth, the sex organs have completely differentiated.

- In hypospadias, the urethra does not connect to the bladder. This is generally surgically corrected at birth.

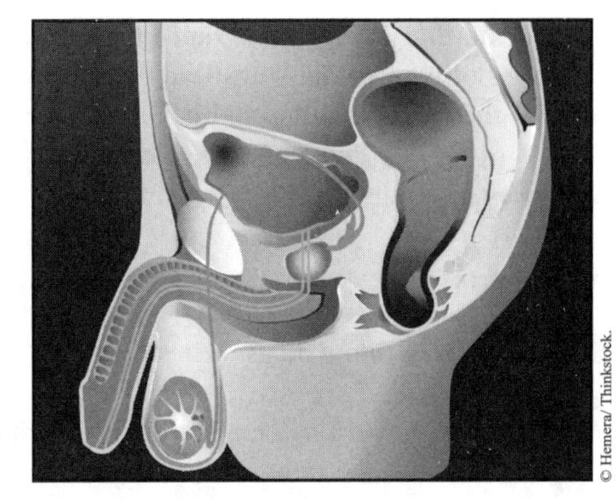


**Cross-section of the human male reproductive system, with parts of the urinary and gastrointestinal systems.**

## Embryology of the internal sex organs

In the undifferentiated stage (5–10 weeks), the gonads, paramesonephric duct, and mesonephric duct are not developed enough to determine the sex of the developing embryo. Between 10 weeks and birth, several developments occur

- After 10 weeks, the testes, epididymis, mesonephric duct (later the spermatocord), and prostate then develop in the male.
- The müllerian duct degenerates in the male; in the female, this duct would have developed into the fallopian tubes.
- Just before birth, the testes descend into the scrotum, and the prostate develops. The sex organs descend in the female as well but remain internal.

## Hermaphroditism

The development to some degree of both sets of sex organs is called hermaphroditism. It is caused by chromosomal abnormality.

## Gross anatomy of the male reproductive system

A series of tubules and ducts lead the sperm and fluids into the epididymis, where the sperm matures and gains motility. From there, the sperm passes through the vas deferens to the ejaculatory duct. Here we examine this pathway in detail.

*Scrotum ("bag")*

- The scrotum holds the testes.
- The superficial fascia (dartos) under the skin can contract and wrinkle the scrotum, drawing it nearer the body for warmth.
- Sperm production is best at about 3° F below body temperature.

*Testes (testicles)*

- The testes develop in the abdomen and descend into the scrotum through the inguinal canal.
- Failure to descend is called cryptorchidism.
    - This condition, if uncorrected, can lead to sterility.
    - It also increases chances of testicular cancer.
    - It can be corrected by hormone injection (human chorionic gonadotropins) or surgery.
- Spermatogenesis is the production of spermatozoa in seminiferous tubules, beginning at puberty.
- The spermatocord migrates from a retroperitoneal position down through the inguinal canal to the testes. It contains the pampiniform plexus, the veins of which can swell and suppress sperm production.
- The cremaster is an involuntary muscle that pulls the testes closer to the body for warmth.

*Ducts*

- The flow of sperm and fluids goes from the seminiferous tubules (tortuous) to the straight tubules.
- Rete (network) testes receive fluid and sperm from the straight tubules.

*Epididymis*

- Efferent ducts lead sperm into epididymis.

- Sperm matures here for about 2 weeks.
- Motility increases.
- The epididymis contains smooth muscle to propel sperm along.

*Vas deferens (ductus deferens)*

- The vas deferens leads from the epididymis to the prostate.
- It enters the inguinal canal and penetrates the peritoneal cavity at the inguinal ring.
- The vas deferens ends in the ejaculatory duct.

*Ejaculatory duct*

- The ejaculatory duct is formed by the junction of the vas deferens and the seminal vesicles.
- It forces sperm out into the urethra during ejaculation.

*Seminal vesicles*

- The seminal vesicles secrete fluids that nurture and protect the spermatozoa.
- The ultimate fluid coating, semen, neutralizes the acidity of the female vagina and protects the sperm.
- Prostaglandins increase sperm motility.

*Prostate gland*

- The prostate gland lies at the base of the bladder at the outlet for the urethra; it acts like a hand-warming muff.

- The ejaculatory duct enters the urethra in the prostate gland.
- Its secretions help sperm motility and longevity.

*Cowper's glands*

- Cowper's glands are paired just below the prostate.
- They secrete a fluid to neutralize the acid environment of the urethra (which is acidic to deter bacterial growth).
- Mucous secretion buffers the sperm against injury during ejaculation.

**Spermatogenesis**

Spermatogenesis consists of the following stages

- Leydig cells (interstitial endocrine cells) secrete testosterone.
- Spermatogenic cells are the stem cells from which all other cells arise. They mature in the basement membrane. They are 2N (or diploid; they have the chromosomes of both the mother and father).
- The spermatogonia (diploid) move upward and become the primary spermatocytes.
- The primary spermatocytes divide into secondary spermatocytes (1N, or haploid; they have only one set of chromosomes).
- Spermatocytes develop into spermatids connected by the cytoplasmic bridge.
- Spermatozoa (mature sperm cells; haploid) gradually move from the seminiferous tubules to the duct lumen as they mature.

- Spermatogenesis takes 2–3 months. Sertoli cells (sustentacular cells) nourish the developing spermatozoa.

**Chromosomes**

There are 23 pairs of chromosomes in the human genome. Chromosome 23 has X and Y variants; these are the sex chromosomes.

- Spermatogonium divide in a process called *mitosis* into daughter cells (diploid) and primary spermatocytes (diploid).

- DNA replication and random crossing over (between chromosome pairs) mix chromosomal material to create a new set of chromosomes.

- In meiosis, primary spermatocytes divide into secondary spermatocytes, each with only one set of chromosomes (haploid).

- Extra chromosomal material in cells always causes problems in embryo development.

**Sperm cells**

- The sperm cell head is pointed to penetrate the egg.

- The nucleus contains one set of chromosomes.

- The cell is also full of mitochondria.

- The tail (flagellum) whips back and forth to propel the sperm out of the body.

- Motility increases as sperm cells mature.

- Individual sperm have little chance of survival

- Each ejaculation contains a million or more spermatozoa.

- No single sperm contains enough of the enzyme necessary to penetrate the egg; thus, many sperm can pool this enzyme.

**The penis**

The penis functions as a conduit for urination and ejaculation of semen. The main components of the shaft are:

- Paired corpora cavernosa, which are spongy collections of pockets that fill with blood during erection.

- A single ventral corpus spongiosum.

- The urethra, which runs through the corpus spongiosum.

- Arteries, veins, and nerves.

- The prepuce, or foreskin, which is sometimes surgically removed in infancy; the benefits versus risks of circumcision are still hotly debated.

*Erection*

A parasympathetic neural response triggers erection as a reation to any of a large number of stimuli (sight, sound, touch, imagination).

The basic processes of erection are as follows:

- Arterial inflow increases.

- Venous outflow is decreased by pressure from arterial expansion.

*Ejaculation*

- Ejaculation is mediated by the sympathetic nervous system.

- The urinary sphincter closes at the base of the bladder, preventing the flow of semen into the bladder and urine into the ejaculate.

- All accessory sexual organs (ductus deferens, seminal vesicles, prostate, and ejaculatory ducts) contract their smooth muscles to expel semen.

- The ischiocavernosus muscle at the base of the penis contracts.

*Semen*

- Semen contains spermatozoa and seminal fluid from the accessory glands.

- Average semen volume is 1 teaspoon (5 ml), which contains about 100 million sperm cells.

- Seminal fluid is bacteriocidal and alkaline.

- It clots initially, then liquefies in 20 minutes.

- Only one sperm is needed to fertilize an ovum.

**Prostate pathology**

- Benign prostate hypertrophy (BPH) tends to close the urethra, causing a number of urinary dysfunctions.

- Transurethral resection of the prostate involves cutting the prostate into small chunks removed through the urethra. This can cause ejaculation into the bladder.

- Cancerous prostates can be approached from the pubic bone or the penis and thus removed, but either surgical approach may cause postoperative problems. ■

## Questions to Consider

1. Why do the testes need to be outside the body cavity in the scrotum? What structure(s) regulate or affect the distance from the body?

2. Describe the pathway taken by sperm from the testes to the urethra.

# Reproductive System—Female

## Lecture 26

**Today we're going to continue our look at the reproductive system by examining the anatomy and the physiology of the female reproductive system. And we'll take a look at both the physiology of menstruation, as well as the physiology of reproduction.**

We begin our study of the female reproductive system by reviewing the anatomy of the external female genitalia, the vagina, the uterus, the fallopian tubes, and the ovaries. Next we consider the physiology of the menstrual cycle, fertilization, and early pregnancy. Finally, we examine the anatomy and physiology of the breast, the various risk factors for breast cancer, and its treatments.

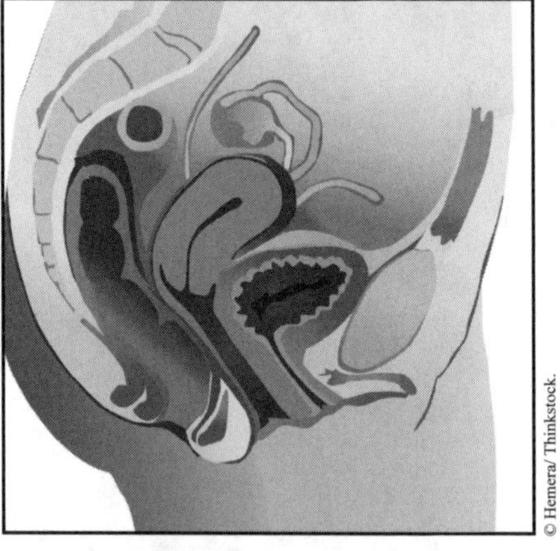



Cross-section of the human female reproductive system, with parts of the urinary and gastrointestinal systems.

### External genitalia

The vulva (volvere, "to wrap around") is also also called the pudendum, and it consists of several parts:

- The prepuce covers the clitoris. Its analog in the male is the foreskin.
- The clitoris is sensitive to stimulation. Its analog in the male is the glans.
- The labia majora ("large lips") are external skin folds; the analog in the male is the scrotal skin and hair.

- The labia minora ("small lips") secrete oil.
- The vestibule is located between the labia minora. It comprises:
  - The hymen.
  - The vaginal orifice.
  - The external urethral orifice.
  - Skene's (paraurethral) glands, which secrete mucous. Their analog in the male is the prostate.
  - Bartholin's (greater vestibular) glands, which secrete mucous during intercourse. Their analog in the male are the Cowper's glands.

The remaining external genital structure is the mons pubis, a deposition of fat that cushions the pubic bone.

**Internal genitalia**
The urinary bladder is a wholly extraperitoneal organ that resides in the vesico-uterine pouch. All female reproductive organs, however, are intraperitoneal.

- The cervix is extraperitoneal (intra-vaginal) and provides easy clinical access to the uterus for examination or removal.
- The rectum and bladder are separated in the female by the uterus. Fistulas can form between the bladder and vagina or the rectum and vagina, leading to inflammation and infection.
- The vermiform appendix lies almost directly above one ovary.
- Pelvic inflammatory disease (tubo-ovarian abscess) causes infections of the tubes and ovaries that can be confused with appendicitis.

*Vagina*

- The vagina is the conduit for spermatozoa during intercourse.
- It functions as the birth canal during delivery.
- It is a potential space, with great elasticity.
- Its walls are composed of smooth muscle.
- It has an acid environment to inhibit the growth of bacteria.
- The cervix of the uterus protrudes into the vaginal vault.

*Uterus*

- The uterus is the conduit for sperm to reach the ovum in the fallopian tube.
- It is the cradle for development of the fertilized ovum (zygote).
- It has the following three anatomic sections:
    - *Fundus*—the superior portion.
    - *Corpus*—the body and midsection.
    - *Cervix*—the "neck" protruding into the vagina.
    - *Cervical os*—the opening of the cervix. There is a mucous plug in the cervical os except during the fertile period.
- The endometrium is the lining of the uterus that covers the muscles.
- The myometrium is the muscular lining of the uterus.

*Uterine pathology*

- Uterine cancer is cancer of the endometrium.
- *Fibromas* (*leiomyomas*) are benign tumors of the uterine wall that cause problems because of their size.
- *Uterine prolapse* is excessive downward movement of the uterus, generally because of age.
- Cervical cancer (*human papilloma virus*) is a common sexually transmitted viral disease.

*Fallopian tubes*

- The active conduit for the ovum to reach the uterus.
- The site of fertilization of the ovum about 24 hours after ovulation.
- It has the following three anatomic parts:
    - Infundibulum—a *fimbriated* ("fingerlike") opening that swirls to suck the ovum into the tube when expelled from the ovary into the peritoneal cavity.
    - Ampulla—the mid-portion and widened part.
    - Isthmus—a narrowed portion leading to the uterus.
- Smooth muscle and cilia propel the ovum distally.
- The fertilized ovum spends about 1 week in the tube in transit.
- In a scarred or damaged tube, pregnancy can take place *in* the tube (ectopic pregnancy), which can lead to tubal rupture and life-threatening bleeding.

*Ovaries*

The ovaries bear all the precursor ova, or eggs.

- Primordial follicles develop into primary follicles that contain oocytes at various stages of development.
- Follicular cells nurture the developing oocyte (estrogen).
- The graafian follicle is the mature follicle at the last stage of oocyte development from which the mature ovum emerges.
- Graafian follicles rupture and release eggs into the fallopian tube. This can sometimes also release drops of blood into the peritoneal cavity, causing *mittelschmerz* (abdominal pain in the middle of the menstrual cycle) for a few hours.
- *Corpus luteum* ("yellow body") remains after ovulation and is the source of estrogen, progesterone, relaxin, and inhibin.
- *Corpus albicans* ("white body") is the remains of degenerating follicle.

**Physiology of menstruation**

- The word "menstruation" is derived from *menses* ("month"), the approximate average time of the cycle.
- Menstruation occurs in phases. The cycle begins with the first day of bleeding.
- The menstrual phase lasts about 5 days, during which time various processes take place:
    - The uterus is the main source of menstrual bleeding.
    - Several secondary follicles begin to mature in both ovaries.

- The preovulatory phase has the most variable duration, between the end of menstruation and ovulation.
  - Ther ovaries are in the follicular phase:
    - FSH (follicle-stimulating hormone) stimulates continued growth of secondary follicles.
    - Only one follicle continues to develop and becomes a graafian follicle.
    - The graafian follicle migrates to the surface of the ovary.
    - The follicle ruptures into the peritoneum. LH (lutenizing hormone) increases at this time.
  - The uterus is in the proliferative phase.
    - Endometrium is repaired and arterial supply increases under estrogen stimulation.
    - This is preparation for arrival of fertilized ovum.
- The next phase is ovulation.
  - Rupture of graafian follicle releases the ovum. LH peaks just prior to ovulation; home tests depend on this to predict ovulation 24 hours later.
  - The uterus enters the secretory phase. First estrogen, then estrogen and progesterone, continue to prepare the endometrium to receive the fertilized ovum. Arterial supply is very rich.
- The postovulatory phase lasts 14 days and is fairly consistent. Estrogen and progesterone decline and cause ischemia to the endometrium, which dies and sloughs off.

Birth control pills mimic pregnancy by maintaining high levels of estrogen and progesterone, which signal the uterus not to release any eggs. This prevents superfecundation (multiple fertilized eggs), which can lead to spontaneous abortion in humans.

**Physiology of fertilization and early pregnancy**

- Fewer than 1% of the sperm reach the ovum.
- Fertilization occurs in the fallopian tube 12–24 hours after ovulation.
- The theoretical window for conception is 36 hours.
  - Sperm are viable for 48 hours after ejaculation.
  - The ovum is viable for 24 hours after ovulation.
- Cleavage of the fertilized egg produces a two-celled conceptum and then, by the fourth day, the morula, which has hundreds of cells.
- The blastocyst, a spear of cells with a hollow center, develops on the fifth day; by the end of the first week of pregnancy, it has implanted itself into the uterine wall.
- If the uterine wall is not ready for the blastocyst, the cell mass will be rejected through a normal menstrual process (missed abortion).
- The blastocyst and endometrium develop two-layer walls around the seventh day.
- The implantation process continues to develop layers of tissue and a separate circulatory system for the embryo.
- The developing placenta grows a network of capillaries that will eventually exchange blood between the mother and embryo.

- In early development, the placenta exchanges only gases, nutrients, and waste because of possible differences in blood type between the mother and embryo.

- Toward the end of pregnancy, the placenta begins to age and can mix the different blood types. If the blood types are incompatible, problems such as neonatal jaundice can occur.

**The breast**

The breast is technically an organ of reproduction because of its role in nurturing the newborn.

- The breasts are also called *mammary glands*.

- They are actually modified sweat (*apocrine*) glands capable of secreting milk.

- The basic structural unit of the breast is the lobule.

    - The lobule is composed of alveoli lined with secretory cells.

    - Several lobules make up a lobe.

    - There are 15–20 lobes in each breast.

    - The lobes drain into ducts lined with epithelium. The duct system converges into 6–8 terminal ducts in the nipple.

- The nipple is surrounded by areola, which contains smooth muscles that eject the milk.

- Ducts empty through the nipple and areola.

- The breast contains ligaments that support it and is divided into primarily breast tissue and fat. The amount of fat increases as the body ages.

- The male breast is similar to the female breast but lacks the ability to secrete milk.

- Men can get ductal breast cancer, whereas women get both lobular and ductal cancers. Because of differences in hormones, the male-to-female breast cancer ratio is approximately 1:100.

**Physiology and pathology of the breast**

- Milk letdown (lactation).

    - Lactation is controlled by oxytocin from the posterior pituitary gland.

    - Prolactin, estrogen, and progesterone are involved in milk production.

- Breast cancer.

    - Breast cancer is the number one cancer in women, but it is highly curable if detected early.

    - It also occurs in men with a similar clinical course.

    - Prolonged exposure to estrogen is the primary cause of breast cancer. Factors that cause prolonged exposure in women include:

        - Cancer in the opposite breast.

        - Nulliparity (no childbirths).

        - Early menarche (early onset of menstruation).

        - Late menopause.

- Family history, especially a first-degree relative with premenopausal bilateral breast cancer. There is less danger in the case of a first-degree relative with postmenopausal unilateral breast cancer.

- Radiation.

- Possibly, diet and obesity/animal fat intake (debated).

○ Treatment includes surgery, radiation, chemotherapy, or all of the above. ■

## Questions to Consider

**1.** What occurs during the preovulatory phase of the menstrual cycle? What is considered day 1?

**2.** At what stage is there rupture of the corpus luteum? What occurs at this time?

# Reproductive System—Physiology of Genetic Inheritance

## Lecture 27

**We, decades ago, always used to talk about genes. We knew that genetic information was in the genes of a cell. That was about the best we could do. Later we knew that the chromosomes ... was where all those genes were. ... That was about as good as we could get. Then in the decades of the middle of the century, we learned that this was made up of a substance called DNA. ... We are down to the real nitty gritty of where the genetic code is.**

This concluding lecture on the reproductive system examines the physiology of genetic inheritance. It begins by identifying the differences between DNA in somatic and germ cells and between genetic and inherited changes in cell DNA.

- At the moment of fertilization, one sperm enters the egg and a calcium-mediated process keeps the other sperm cells out.

- *Mitosis* is the division of cells into daughter cells with full sets of chromosomes. This can be seen in a microscope.

    - Normal somatic cells, even very active ones, divide so infrequently that if mitosis can be seen in a somatic cell, it is considered evidence of cancer.

    - When daughter cells divide for the first time, they can separate completely, producing two organisms with the same DNA (twins).

    - Further cell division creates a blastocyst that is implanted in the uterine wall.

- Differentiation.
  - Blastocyst cells are *totipotential* or *pluripotential*, meaning that they can become any type of cell.
  - Totipotential cells will eventually differentiate into specific types of functional cells, then continue to specialize during development.
  - Cell differentiation is irreversible.
  - The embryonic neural crest is the beginning of the vertebra and nervous tissue.
- The developing embryo displaces all organs in the mother's abdomen, making abdominal disease or infection during pregnancy hard to treat.

**In utero tests**

- Amniocentesis is the withdrawal of amniotic fluid from the amniotic sac surrounding the embryo. Tests can determine abnormalities, such as trisomy 21 (Down syndrome) and meningomyelocele (failure of the neural crest to close).
- Chorionic villi sampling uses small pieces of the placenta and can be done earlier than amniocentesis (8–10 weeks versus 14–16 weeks).
- Intrauterine surgery can correct certain problems if performed early enough.

**Genetics**

- Somatic cells (normal body tissue) have two sets of chromosomes (*diploid*). Germ cells have one set (*haploid*).

- Watson and Crick discovered the famous double helix structure of DNA. Four different components (C, G, A, and T) produce the code our body uses to construct itself.

- Most DNA has no meaning; as little as 10% has functional use.

- Differences between the DNA in somatic cells and germ cells.

  - In somatic cells, DNA is diploid, each cell carries the entire DNA complement, and mutations in the cell are transmitted only along that particular cell line during replication.

  - In germ cells, DNA is haploid. This DNA represents only one side of the genetic complement (male or female), and mutations will show up in every cell of the offspring.

**Differences between "genetic" and "inherited"**

- *Genetic* refers to changes in the DNA of any cell. They can be somatic or germ line changes.

- *Inherited* refers to changes in the DNA that are passed on through the germ line. All the body's cells will reflect this change.

- *Congenital* means present at birth. Therefore, these changes can be inherited, as well as genetic (Down's syndrome versus fetal alcohol syndrome).

**Principles of inheritance**

- Genotype versus phenotype.

  - Genes are locations in the nuclear DNA that carry instructions for body processes. Gene proteins carry the instructions.

  - *Genotype* describes the genetic makeup of the cell regarding its complement of genes.

- *Phenotype* describes how that complement is expressed by the body.
- Pairs of chromosomes are called *homologues*.
- Genes at the same location on a chromosome (for the same trait) are *aleles*.
  - Dominant alleles express the trait.
  - Recessive alleles are overridden by the dominant one.
  - Two recessives in a cell, without the dominant to override them, will express the recessive trait (e.g., tasters versus non-tasters).
  - *Incomplete dominance* is when a compromise is expressed in a cell with mixed dominant and recessive genes (sickle cell anemia, sickle trait, normal).
  - Most harmful genes are recessive, while a few are dominant (Huntington's disease).

**Homozygous versus heterozygous**

- If a cell has similar alleles at a specific locus, it is *homozygous* for that trait (both loci contain sickle cell anemia genes).
- If the genes at that locus are different, it is *heterozygous* (one gene for sickle cell anemia and one normal).
- A *Punnett square* is a matrix that lists all possible allele variations in children given parents with specific alleles.
  - Phenylketonuria (PKU), for example, is an inability to convert phenylalanine to tyrosine, producing phenyl ketones in the

blood. A Punnett square shows that, out of four children with parents who carry phenylketonuria but do not have the disease:

- One will neither have nor carry the disease.
- Two will carry but not have the disease.
- One will both have and carry the disease.
- This distribution is very common.

**Sex chromosomes**

- Sex is determined by combinations of X and Y chromosomes.
- Females have only X chromosomes; males have an X and a Y.
- The XX combination is female; the XY combination is male.
- Extra genetic material (because of two X's in the female) is taken care of by X chromosome inactivation after conception, leaving only one active X component in the female.
- Female and male development is the same until about the seventh week after conception.

**Incomplete penetrance**

Incomplete penetrance is one result of a heterozygous gene pair. Consider a person who is heterozygous for the sickle cell anemia gene:

- Sickle cell anemia is selected *for* along the equator by malaria. Of four children whose parents are only carriers of sickle cell anemia, statistically speaking:
  - One will neither carry nor have the disease.

- Two will carry the disease and have a mild form (sickle trait). This is called *incomplete penetrance.*

- One will both carry and have sickle cell anemia.

**Sex-linked traits**

- Genes on the X chromosome will be transmitted by the mother in higher percentages to sons than to daughters. These include red-green color blindness and hemophilia.

- The Y chromosome gene locus (sex-determining region) starts male differentiation.

    - Because there is no female contribution of Y chromosome material, there is no mixing.

    - The Y chromosome has been used recently to trace back genetic origins of various populations.

**Polygenic inheritance**

- Some traits (e.g., hair, skin, and eye color) involve multiple alleles.

- This leads to a much greater range of possible variations than with one allele per trait.

**Developmental abnormalities**

- Sex-related diseases:

    - In Turner's syndrome, no second sex chromosome exists in females, leading to sterility and small size.

- In Klinefelter's syndrome, males receive an extra X chromosome, leading to sterility, slight mental retardation, and breast enlargement.
- Females who receive three X chromosomes become mentally retarded, sterile, and underdeveloped.
- Amniocentesis and chorionic villi sampling can detect these diseases but carry a risk of spontaneous abortion.
- There are also the ethical considerations related to the purpose of detecting incurable diseases.

- Non-disjunction:
  - In non-disjunction, chromosomes fail to separate properly during meiosis, leaving an abnormal number (too many in one cell and too few in another).
  - The Philadelphia chromosome, which causes leukemia, is a non-disjunction.
- Cardiovascular anomalies:
  - Patent ductus arteriosus: Aortic blood flows into the pulmonary artery, increasing pulmonary pressure and overloading both ventricles.
  - Interatrial septal defect.
    - This condition results from failure of the foramen ovale to close after birth.
    - Blood flows from the left to the right atrium without going into the systemic circulation.

        - This increases the load on the pulmonary side of the circulation.

        - It can be repaired by surgery.

    - Interventricular septal defect.

        - This condition results from failure of closure of the ventricular septum.

        - It leads to mixing of oxygenated left-sided blood with the deoxygenated right-sided blood.

        - This is inefficient and causes overloading of the myocardium.

    - Tetralogy of Fallot (a set of four congenital defects):

        - Overriding of the aorta to both sides.

        - Pulmonic stenosis.

        - Interatrial septal defect.

        - Right ventricular hypertrophy ("blue baby"; deoxygenated blood mixes with systemic circulation, resulting in very little pulmonary blood supply).

        - Surgical correction involves coarctation of the aorta:

            - This condition involves stenosis of the aorta, usually in the arch.

            - It leads to high blood pressure in the arms and head, low blood pressure in the lower body and legs, and hypertrophy of the left heart muscle. ■

## Questions to Consider

1. What is the purpose of X chromosome inactivation? In which sex does it occur?

2. A father has PKU and the mother does not. Together, they have a child who also has PKU. What is the father's genotype?

# Musculoskeletal System—Physiology and Physics of the Muscles

## Lecture 28

**The basic functions of the muscular system ... is to first of all provide stability and postural tone for the body. ... We've got to be able to maintain our posture in space. After that, we need to have purposeful movement, so those muscles are going to have to take this posture in space.**

This is the first of three lectures on the musculoskeletal system. In this lecture, we will examine the physiology and physics of the muscles. Basic functions of the muscles are to provide stability and postural tone, to allow purposeful movement, to regulate internal organ movement and volume, and to generate heat.

Muscle mass constitutes 50% of body weight. The body loses heat in relation to the square area of skin, but heat is generated in relation to the cubic volume of mass, much of which is provided by the muscles. Shivering is the body's attempt to generate heat through rapid muscle contraction.

Muscles can only actively shorten (contract); thus, all motion is accomplished by opposing muscle pairs. Muscles are conductors and respond to electrical stimulation by contracting.

### Muscle types

- Skeletal muscle attaches to bone, skin, fascia, and other muscles. It is similar to heart muscle.
  - Specific nerves stimulate specific muscle fibers.
  - Skeletal muscle is voluntary.

  - Skeletal muscles are striated (striped), with bands of muscle fibers made of *actin* (thin filament) and *myosin* (thick filament).

  - The thin and thick filaments have microscopic spurs that ratchet into each other during muscle contraction.

  - Muscle tissue itself is not particularly strong.

- Smooth muscle makes up the walls of hollow organs, hair follicles, and blood vessels. It mostly regulates the size of intestinal muscles and glands.

  - Smooth muscle is microscopically smooth (not striated), with irregular bundles.

  - It may be innervated by one nerve or multiple nerves, depending on function.

  - It is involuntary.

- Cardiac muscle makes up the walls of the heart.

  - It is microscopically striated, like skeletal muscle.

  - Its striations join together in bundles that allow coordinated action.

  - It is involuntary and autorhythmic. Some cardiac muscles function as built-in pacemakers.

**Basic components and definitions of the muscular system**

- Nerve supply:

  - Muscles are stimulated by motor neurons.

- The cell body is in the spinal cord, with the axon ending in the muscle.

- Muscles are richly supplied with arteries and veins and have intimate contact with a rich capillary network, because of high energy demands.

- Motor unit:

    - The motor unit is composed of a motor neuron and the muscle fibers it supplies.

    - Innervation of the motor neuron may vary from as few as 10 muscle fibers to as many as 2,000.

    - Larger muscle groups each have larger motor units (back muscles, thighs, and so on).

    - Smaller, finer movement muscles have a larger number of motor neurons per muscle (eye muscles, tongue, larynx).

    - When a motor neuron fires, all the muscle fibers in its motor unit fire.

    - Physiology of nerve and muscle conduction in the neuromuscular junction (NMJ):

        - Nerves connect with their target muscle at the NMJ.

        - The *synapse* is the place where the nerve meets the muscle or another nerve.

        - The nerve ending is separated from the muscle by the synaptic gap, a microscopic physical space.

        - In place of direct electrical stimulation of the muscle by the nerve, an intermediary is needed: the neurotransmitter.

    - Each neuron axon has an axon terminal that ends in synaptic bulbs. Each bulb contains synaptic vesicles.

    - The synaptic vesicles contain many molecules of the neurotransmitter.

    - Acetylcholine (Ach) is the most common neurotransmitter.

    - The motor end plate is the region on the muscle side of the synaptic gap in proximity to the axon terminal bulb.

    - The motor end plate contains the Ach receptors, which when touched by the Ach molecule, fire the muscle to contract.

    - Acetylcholinesterase is an enzyme that breaks down the Ach and ends the contraction stimulus.

    - Because most neuromuscular junctions are located in the middle of the muscle fiber, the wave spreads from the middle outward toward the end, and the muscle makes a smooth contraction.

- Energy

    - Adenosine triphosphate (ATP) and diphosphate (ADP) store chemical potential energy in muscle cells.

    - ATP is converted to kinetic energy and ADP, to perform work. *Energy* is the ability to perform work; *work* is moving any object any distance.

    - ADP is then converted to ATP for another round of work.

**Types of skeletal muscle**

- Type I fibers: slow oxidative fibers.
    - These fatigue-resistant fibers are also known as *slow-twitch fibers*.
    - They generate ATP aerobically (using oxygen), which is very efficient.
    - They split ATP slowly and contract slowly.
    - They are postural muscles.
- Type IIA fibers: fast oxidative fibers.
    - These somewhat fatigue-resistant fibers are fast-twitch A.
    - They generate ATP aerobically.
    - They split ATP rapidly and contract rapidly.
    - They are less fatigue-resistant than Type I fibers.
    - They are typical of sprinter's muscles in track.
- Type IIB fibers: fast glycolytic fibers.
    - These fatigable fibers are fast-twitch B.
    - These fibers generate ATP anaerobically (without oxygen).
    - They split ATP rapidly but not at a steady rate.
    - They are typical of large muscles in the arms.

## Characteristics of the muscle types

Most muscle groups are mixtures of the Type I, IIA, and IIB categories. Muscles vary in the proportions of these types.

- Postural muscles are mainly Type I.
- Because muscles of the shoulders and arms are used in short bursts rather than continuously, they need more fibers of Type IIB.
- Large muscle groups of the legs need postural, walking, and sprinting fibers, so they have all three groups.
- Types can be changed by different exercise patterns. Going from sprint training to steady-state training, as in long-distance running, will convert the proportions of fiber types.

## Muscle contraction

Muscle contraction is an "all or none" phenomenon.

- A muscle fiber that is fired contracts to its whole length or not at all.
- The force of contraction is generally maximal.
- In pathological states, the brain may fire constantly, randomly, or inappropriately, causing muscle contraction.
    - Grand mal epileptic seizures involve the entire brain firing at once, and patients tend to assume characteristic postures that depend on the strength of certain muscles.
    - Spastic paralysis, where all arm muscles fire at once, causes the hand to bend down, because the adductor, biceps, and pronator muscles are stronger than their opposing muscles.

- The force of contraction depends on stimulations per second (frequency), muscle length prior to contraction, and number of motor units stimulated.

- *Muscle tone* refers to continuous involuntary contractions of small numbers of motor units that give the muscle its firmness even at rest.

- *Isotonic contraction* puts the same tension on all muscles.

- *Isometric contraction* keeps the length of the muscle the same throughout, which constantly changes the force.

- In *concentric motion* (positive repetition), the angle between the two limbs decreases.

- In *eccentric motion* (negative repetition), the angle between the two limbs increases.

    - Eccentric motion damages muscle more than concentric motion.

    - This is useful for weightlifters, because the point of exercise is to slightly damage muscle, which will receive added mass and strength when it is rebuilt.

**Exercise**

- Regular, repeated, low-level exercise, such as jogging, is aerobic—the body can meet its oxygen demands.

- Anaerobic exercise, such as weightlifting, builds mass by damaging muscle but can, of course, lead to overdamage.

- A healthy exercise routine mixes aerobic and anaerobic exercise.

- Interval training combines aerobic and anaerobic exercise in the same activity by varying the intensity.

- Exercises involving gravity are beneficial for the bones and bone density. On the other hand, swimming allows patients with joint and bone problems to exercise without undue stress.

- Muscle at rest burns more calories than fat at rest; thus, adding muscle mass helps control weight.

- Contrary to old beliefs, regular exercise builds muscle mass and improves health even in the very elderly. ■

## Questions to Consider

1. Describe the chemical reaction at the neuromuscular junction across the synapse.

2. Distinguish between isotonic and isometric contraction.

# Musculoskeletal System—Anatomy of the Muscles

## Lecture 29

**There are 700 skeletal muscles whose function is to produce movement and stability and act with the bones as a system of leverage. So it's important, now that we've had a look at the microstructure of muscles, to see how they fire and what their physiology is, to get a little bit of an understanding of the physics.**

In this lecture, we examine the anatomy of the muscles and their operation as levers in conjunction with the bones (most human muscles operate as third-class levers). Next, we review the various names of the muscles, as indicated by their size, shape, orientation of their fibers, mechanical action, number of origins, origin and insertion points, function, and location.

- The human body has nearly 700 skeletal muscles.
- These produce movement and stability by acting as levers in conjunction with the bones of the skeletal system.

**Basic physics of simple levers**

- Levers can increase strength of movement, speed of movement, or range of movement, but they cannot increase all three.
  - If strength is increased by mechanical advantage, range of motion is diminished.
  - If range of motion is increased by mechanical advantage, strength is diminished.
- A first-class lever is the fulcrum is between force and resistance.
  - If force is nearer the fulcrum, speed is gained.
  - If load is nearer the fulcrum, force is gained.

- There are very few first-class levers in the body. One example is in the neck: The head rests on the atlas bone, pulled erect by the muscles of the back of the neck.

- In a second-class lever, load is located between the fulcrum and the force.
  - Increase in strength comes at the expense of range of movement.
  - There are very few second-class levers in the body.
  - Think of the body as being like a wheelbarrow: Standing on tiptoes, the toe is the fulcrum, the body is the load, and the gastrocnemius (calf muscle) is the force.

- In a third-class lever, force is located between the fulcrum and the load.
  - Range of movement and speed are gained at the expense of strength.
  - Most muscle groups of the body are third-class levers. For example, flexion of the forearm (i.e, performing biceps curls): the elbow is the fulcrum, the weight in hand is the load, and biceps muscle is the force.

**Basic terminology**

- *Proximal*: near the top of the organ or muscle.
- *Distal*: away from the top of the organ or muscle.
- *Origin*: attachment of the muscle to the more stationary bone or structure.

- *Insertion*: attachment of the tendon to the moving bone or structure.
- *Tendon*: dense connective tissue that connects muscle to bone.
- *Tubercule*: a thickened area of a bone where a tendon attaches.
- *Belly*: the middle portion of a muscle.
- *Ligament*: tissue that attaches bone to bone to stabilize joints.

**Names of muscles by function**

- Muscle names based on size:
    - *Maximus* (or magnus): largest of a group (i.e., gluteus maximus, adductor magnus).
    - *Minimus*: smallest of a group (i.e., gluteus minimus).
    - *Longus*: longest of a group (i.e., flexor pollicis longus).
    - *Brevis*: shortest of a group (i.e., flexor pollicis brevis).
    - *Latissimus*: widest of a group (i.e., latissimus dorsi).
- Muscle names based on shape:
    - *Trapezius*: shaped like a trapezoid.
    - *Deltoid*: triangular.
    - *Serratus*: saw-toothed (i.e., serratus anterior).
    - *Platysma*: flat.

- Muscle names based on orientation of fibers:
    - *Rectus*: parallel to the midline of the body (i.e., rectus femoris, rectus abdominis).
    - *Oblique*: diagonal (i.e., internal oblique of the abdomen).
    - *Transverse*: perpendicular to the midline of the body (i.e., transversus abdominis).
- Muscle names based on mechanical action of the muscle.
    - *Flexor*: decreases the angle at a joint (i.e., flexor pollicis longus).
    - *Extensor*: increases the angle at a joint (i.e., extensor pollicis longus).
    - *Pronator*: turns downward or backward (i.e., pronator teres).
    - *Supinator*: turns upward or anterior (i.e., supinator of the forearm).
    - *Levator*: lifts upward (i.e., levator ani).
    - *Depressor*: pushes or moves downward (i.e., depressor labii inferioris).
    - *Abductor*: moves bone away from midline (i.e., abductor magnus).
    - *Adductor*: moves bone toward the midline (i.e., adductor brevis).
    - *Tensor*: makes rigid (i.e., tensor fascia lata).

- *Sphincter*: closes an opening (i.e., anal sphincter).
- *Rotator*: produces a circular movement around a joint (i.e., rotator cuff of the shoulder).

- Muscle names based on number of origins:
  - *Biceps*: two origins (i.e., biceps femoris).
  - *Triceps*: three origins (i.e., triceps brachii).
  - *Quadriceps*: four origins (i.e., quadriceps femoris).
- Names based on origin and insertion.
  - The sternocleidomastoid muscle originates from the sternum and clavicle.
  - It inserts into the mastoid bone.
- Muscle names based on function:
  - *Risorius*: laughing.
  - *Masseter*: chewing.
- Muscle names based on location:
  - *Temporalis*: located on the temporal bone.
  - *Zygomaticus*: located on the zygoma.
  - *Sartorius* (from "tailor"): located in the thigh and knee; this muscle can be used to cross the legs in the manner of a tailor sewing, hence the name.

### Other connective tissue

- *Fascia* surrounds most skeletal muscles.
  - It holds the major muscle fibers in a group together.
  - It supports vessels, nerves, and the lymphatic channels.
- *Aponeurosis* is a tendon-like structure that is broadly spread out to attach muscle to skin, fascia, or other muscle.
- A cage of tendons running from the thigh to the leg protects the knee from instability and damage. Damage to these tendons requires extensive physical therapy; otherwise, the knee will never recover its full strength.
- One of the largest nerves in the body, the sciatic nerve, runs under the buttocks and into the leg. The "need" to stretch this nerve is a popular misconception; stretching a nerve will only damage it. ■

## Questions to Consider

1. Describe a first-class lever. What muscle group and bone falls into this category?
2. Why are third-class levers preferred in nature?

# Musculoskeletal System—Bones

## Lecture 30

**The primary functions of the skeletal system are very much related to the muscular system. ... It's the latticework of the structure upon which we can build the muscles for movement, and it works in conjunction with the muscular system, again to provide the levers that are going to be moved and which will provide stability for position and space.**

In this final lecture on the musculoskeletal system, we examine the divisions and main functions of the skeletal system, which works in parallel with the muscular system to provide the body with support and motion.

- There are 206 named bones, ranging from enormous bones, such as the pelvis and the femur, to the tiny ossicles of the inner ear.
- The *axial bones* are lined up vertically for support. They are parallel to the center of gravity.
- The *appendicular bones* are mainly limbs and girdles that attach limbs to the axial system.

The primary functions of the skeletal system are:

- Movement, in conjunction with muscular attachments.
- Support of the body.
- Physical protection of internal organs.
- Production of red blood cells, white blood cells, platelets, and macrophages (red marrow).
- Mineral storage (calcium, phosphorous, and magnesium).

- Lipid storage for emergencies (yellow marrow).

**Gross anatomy**

- General kinds of bones:
    - *Flat bones* parallel the surfaces of the body and have a protective function. There is a broad attachment surface.
    - *Long bones* are weight-bearing, curved, and strong. They are both compact and spongy. The limbs are composed of long bones.
    - *Short bones* are compact on the surface, with spongy centers, and are cuboidal. Wrist and ankle bones are short bones.
    - All the rest are *irregular bones*, which have various functions. Examples include the vertebrae and facial bones.
- *Sesamoid bones* are embedded in tendons to change the direction of movement, like a pulley; the patella (kneecap) is an example.
- The *diaphysis* (*dia*, "through," *physis*, "growth") is the long section of the bone.
- The *metaphysis* (*meta,* "after") is the intermediate area between the diaphysis and the epiphysis. It is where growth occurs in young bones.
- The *epiphysis* (*epi,* "above") is the end of the bone.
- *Articular cartilage* lines bone-to-bone joints, allowing frictionless movement.
- *Periosteum* is the fibrous covering of the bone. It is the source of vessels and nerves. Ruptures of the periosteum are the source of bone pain.

- The *marrow cavity* is the hollow center of the bone in the diaphysis.

**Blood supply to the bones**

- The nutrient artery generally enters through the diaphysis.
- The epiphyseal artery enters through the ends of the bone.
- The periosteal artery nourishes the periosteum. There are many points of entry.

**Surface markings of bones**

- *Tuberosity*: roughened, rounded knob.
- *Tubercle*: smaller knob.
- *Spine* or *spinous process*: slender projection.
- *Trochanter*: large projection of the femur.
- *Condyle*: large knob or rounded surface.
- *Epicondyle*: smaller prominence above the condyle.
- *Facet*: flattened surface of a joint attachment.
- *Crest*: ridge.
- *Sinus*: hollow space.
- *Meatus*: tunnel or canal.
- *Fossa*: depression.
- *Foramen*: hole or opening.

- *Fissure*: cleft.

**Bone development**

- Bones begin in the embryo as cartilage that gradually calcifies.
- Hollow areas form and spread up and down the bones.
- Nutrient arteries and matrices (series of molecules without calcium) develop.
- Finally, the diaphysis and metaphysis become truly separate.
    - The epiphyseal plate is the border between the end of the metaphysis and the beginning of an epiphysis. It is the only place where growth occurs, stimulated by growth hormone.
    - In children, the epiphyseal plates are open to allow growth; in adults, they are closed.
        - An excess of growth hormone in children leads to a giant but normally proportioned body.
        - An excess of growth hormone in adults will only thicken bones, because they cannot grow, leading to thick and exaggerated features.
        - If an open plate is injured, its bone will stop growing.

**Microscopic anatomy**

- *Osteoprogenitor cells* are multipotential skeletal cells.
- *Osteoblasts* are bone-forming cells.
- *Osteocytes* are mature bone cells that become trapped at maturity in a matrix.

- *Osteoclasts* are bone-reabsorbing cells that destroy old bone.

- *Collagen* is soft, strong connective tissue that supports and reinforces the mineralized matrix.

**Fractures—healing stages**

- Fractures commonly tear blood vessels, producing a hematoma that can be used to anesthetize the periosteum.

- After several weeks, the periosteum is beginning to heal and lay down scar tissue. Trabecular bone has begun to grow over the break.

- After several months, new blood vessels have formed (angiogenesis), and further callus growth and new cell deposition have significantly strengthened the bone.

- Over the next months or years, the bone shape returns to normal as osteoclasts absorb extra cells and osteoblasts generate new cells and bone. The bone will be thickened somewhat at the fracture site.

- Fully healed fractures in children are indistinguishable from the original bone because the growth plates are open. However, multiple fractures in various stages of healing are a strong indicator of child abuse.

**Fractures—terms commonly used**

- *Reduction* is the reestablishment of the normal position of fractured bones and dislocations.

- *Dislocation* is the displacement of a bone from a joint.

- *Fracture* is a break in the bone itself.

- *Internal fixation* is the surgical placement of steel material into the bone to hold it in place. This can be done through a surgical incision or through fluoroscopic guidance from without.

**Fractures—different kinds**

- A *Colles' fracture* of the radius is a very common result of a fall.

- In a *Potts fracture*, some incident, such as a misstep off a curb, turns the ankle, and the ligament tears off the bone or fractures it.

- A *greenstick fracture* is an incomplete fracture with part of the bone intact. It is common in the soft bones of children. Treatment involves reduction and immobilization.

- In an *impacted fracture*, two fragments are pressed tightly together. Treatment involves immobilization and sometimes repositioning.

- In a *comminuted fracture*, the impact shatters the bone into multiple small fragments. Treatment varies from closed reduction to open surgical repair.

- In an *open* (*compound*) *fracture*, the bone fragment penetrates the skin, leaving the bone open to infection (*osteomyelitis*), which can be dangerous. Treatment involves surgical debridement, reduction, and immobilization.

- A *pathologic fracture* occurs secondary to a weak spot from disease (often malignancy) in the bone. Treatment involves internal (surgical) fixation, because the bone cannot heal itself.

- *Stress fractures* are tiny hairline fractures from repeated insult to a particular bone or group. *Tibia fractures* ("shin splints") in runners and foot fractures ("march fracture") in soldiers are common.

**Joint diseases—arthritis**

- The synovial membranes secrete fluid that lubricates the bones.
- Osteoarthritis is a degenerative disease, resulting from wear and tear on joints.
  - It affects articular cartilage primarily. The disappearance of cartilage causes bones to rub together, causing inflammation, bone spurs, and bone deterioration.
  - It affects large joints first but can spread to any joints.
  - It affects both synovial and weight-bearing joints.
- Rheumatoid arthritis is an autoimmune disease in which patients become allergic to their own joints.
  - It primarily affects synovial membranes; the membranes become thick and inflamed.
  - It primarily affects small, fine joints, such as the fingers.
  - It tends to be bilateral and symmetric.
- Osteoarthritis is far more common than the rheumatoid version, but rheumatoid arthritis is more crippling. ■

## Questions to Consider

1. Name and describe at least four functions of the bones.
2. If the radius and ulnar heads are fractured (Colles' fracture), what joints must be immobilized in a cast to control movement until healing takes place?

# Immune System—Anatomy and Physiology

## Lecture 31

**The immune system: This is the system that sets up surveillance and protection for the recognition and destruction of foreign biological invaders, what I guess we would call in our society today "stranger danger."**

This lecture examines the immune system, which comprises all of the body's mechanisms for defending itself against foreign invaders. This doesn't include chemical invaders such as toxins, which par processed by the liver, Insted, we're talking about living invaders.

- The system is specific, wide-ranging, and has long memory.
- It distinguishes self from non-self with great accuracy.
- It is extremely tolerant to substances encountered during embryologic development.
- It differs from non-specific defenses (e.g., the skin).

### Basic definitions

- *Immunocompetence* is the ability of cells to function as mature cells in the immune system and carry out certain tasks in protecting the body from foreign invasion.
- *Lymphocytes* are white blood cells that destroy foreign invaders.
- *Antigens* are foreign molecules that provoke an immune response from the body by specific geometric arrangements of their molecular surfaces.
- *Antibodies* are molecules that react with specific antigens to kill or neutralize them.

**Components of the system**

- Cell-mediated immunity.
    - T-cells develop from stem cells, which migrate from red bone marrow to the thymus gland.
    - They mature in the thymus.
    - Antigen receptors develop on the surface of the cell.
    - T-cells are activated by T-cell receptors (TCRs) that recognize antigens.
        - TCRs are unique and number in the millions.
        - Costimulators (interleukins or cytokines) are a second type of molecule needed to complete the reaction of the TCR with the antigen. They provide a failsafe backup mechanism.
        - After recognition and costimulation, the T-cell is sensitized and begins to clone millions of T-cells with similar TCRs.
    - T-cells acquire one of two kinds of surface proteins, CD4+ or CD8+, and are called, respectively, T4 or T8 cells. CD4+ and CD8+ regulate the response processes of other cells.
    - Kinds of T-cells.
        - Helper T-cells. These are mainly T4 cells (CD4+).
            - They help activate T8 cells and turn them into killer T-cells.

- They help activate antibody-mediated immunity.
- They produce interleukin-2 to stimulate virtually all immune processes.
- They produce a clone of memory helper T-cells, which last a long time to resist the next invasion of that specific antigen.

- Cytotoxic (killer) T-cells. These are mainly T8 cells (CD8+).
  - They recognize specific foreign invaders (primarily intracellular) and kill them by lysing (breaking apart) invading cells.
  - They require costimulation by interleukins produced by helper T-cells.
  - They require direct contact (the "kiss of death") to kill cells but can kill multiple cells.
  - They secrete perforin, which makes holes in the invader's plasma membrane and allows flooding into the cell.
  - They secrete lymphotoxin, which kills enzyme systems in the invader.
  - They secrete gamma interferon, which stimulates macrophages to eat foreign invaders.
- Suppressor T-cells.
  - These cells cause downregulation of immune response.

    - Very little is known about them.
    - Some cancer cells signal suppressor T-cells to deactivate the immune system before spreading.

- Memory T-cells.
    - They last a long time to mediate a quicker response to the same invader at a later date.
    - They are a subset of the original clone.
    - They may act before any signs of invasion (no clinical symptoms).

**Humoral (antibody-mediated) immunity**

- B-cells from the bone marrow remain in place in the lymph nodes, spleen, and intestinal lymphatic tissue (Peyer's patches).
- They primarily target extracellular agents, such as bacteria.
- They are activated by recognition of a specific antigen.
- They are costimulated by interleukins from T-helper cells.
- They differentiate into plasma cells.
- Plasma cells secrete antigen-specific antibodies at a rate of millions of molecules per cell per second.
- The duration of plasma cell response is up to 1 week until the death of that plasma cell.
- Memory B-cells remain in the system to respond rapidly to the next infection/invasion (accelerated response).

- Each clone of plasma cells from a specific clone of B-cells secretes a specific antibody for a single antigen.

- Antibody action.

    - Antibodies agglutinate and precipitate antigen.

    - They immobilize ciliated or flagellated bacteria.

    - They neutralize toxins from bacteria, such as tetanus. Tetanus toxoid injections artificially stimulate antibody production before exposure.

    - They stop viral entry into cells.

    - They increase phagocytosis by activating complement system and opsonizing (coating) bacteria.

    - Monoclonal antibodies make use of specificity for diagnosis and delivery of chemicals, radiation, and so on.

    - Rabies is a virus that attacks the central nervous system and destroys the brain. It can sometimes be interrupted by antibody response if it is stopped on its way to the brain.

**Natural killer (NK) cells**

- These are lymphocytes that kill a wide range of invaders.

- No prior exposure or sensitization is needed.

- They do not mature in the thymus.

- They do not have surface antigen receptors.

- They reside in the spleen, blood, bone marrow, and lymph nodes.

- They must contact the invader.
- Recognition is probably related to major histocompatibility (MHC) molecules.
    - Major histocompatibility (MHC) antigens are surface markers on all body cells (except red blood cells).
    - They are unique glycoproteins, recognized by the body as "self."
    - Foreign invaders without MHC antigens are seen as "non-self."
    - MHC molecules take fragments of foreign antigen and "present" them to the T-cells for inspection.
    - The T-cell then recognizes the presented fragments as "self" or "non-self" and will attack "non-self" while ignoring "self."
    - Antigen presenting cells (APCs) are located where foreign entry is likely to occur (respiratory tract, skin, gastrointestinal tract, urinary tract, and so on).
    - MHCs are very important in the recognition by the body of transplanted tissues from the donor (e.g., kidney from non-identical twin).

**Inflammation**

- The four signs of inflammation are *rubor* (redness), *tumor* (mass), *calor* (heat), and *dolor* (pain).
- Tissue injury opens the door for microbes to breed.
- Chemotaxis draws macrophages and polymorphonuclear leukocytes (white cells).

- The blood vessels in the area of injury are signaled to increase their flow (producing rubor) and permeability (which causes edema and swelling [tumor]).

**Types of immunity**

- In active immunization, the body mobilizes the immune system in response to a foreign invader.

- Passive immunization is the infusion of active antibodies against a specific disease or invader, either by direct injection or through the womb.

- Specific immunization is the injection of a reduced-strength antigen into the blood to enable the body to then produce its antibodies.

- Nonspecific immunization is the injection of certain antigens that raise the level of immune function throughout the body.

- Adopted immunotherapy is direct injection of, typically, interleukin-2 to stimulate the immune system. ■

## Questions to Consider

1. How do natural killer cells differ from T-cells in their recognition of an enemy?

2. What cells are affected by HIV virus? What is the difference between HIV infection and AIDS?

# The Biology of Human Cancer
## Lecture 32

**In this lecture, we will examine the subject of human cancer. We will see how the fragility of the basic information molecule, DNA, allowed evolution from the primordial chemical "soup" into modern humankind and how the same fragility that allowed us to evolve has saddled us with susceptibility to mutations that can cause cancer.**

There are more than 6–7 million new cases of cancer diagnosed worldwide per year, not including non-melanoma skin cancers. Eight million Americans now alive have had cancer. Of those 8 million, 5 million are considered cured. Cancer is the second-leading cause of death in America, behind heart disease. Obviously, the biology of cancer has a significant impact on many, many lives.

- Cancer is a genetic disease, though less frequently inheritable or congenital.

- DNA is a mutable molecule; that is, it can be changed by a variety of factors. This can be good (in evolution) or bad (in cancer and other genetic defects).

- Cancer is basically a microevolutionary process that occurs at a molecular level.

- Cancer is largely a self-inflicted wound; 80% of cancers are from environmental causes:

    - Chemicals, especially industrial and agricultural.

    - Ionizing radiation, including UV radiation.

    - Asbestos and radon—disturbing the environment.

- Sexually transmitted diseases, such as human papilloma virus and HIV.

**The four characteristics of cancer cells**

- Failure to differentiate.
  - Cancer cells come from stem cells that become injured and prevent differentiation.
  - Because the body has its full complement of neurons at birth, brain cancer comes mostly from cancerous glial or supporting cells that are replaced during life.
- Potential to invade.
  - Normal *benign* ("bringer of good") cells do not pass their basement membrane.
  - *Malign* ("bringer of bad") cells have the potential to invade other parts of the body.
  - For example, in the breast, a malignant tumor may not have broken the basement membrane, but it is capable of doing so.
- Potential to metastasize.
  - Cancer cells can travel to different parts of the body through the blood and start new tumors.
    - *Primary cancer* refers to the location of the primary tumor.
    - *Metastatic cancer* refers to any cancer that has moved from the original site.

- Lethal by design.
    - Benign tumors kill patients only accidentally.
    - Malignant tumors will always metastasize and kill if untreated.

**Cancer terms**

- The ending *–oma* almost always refers to benign tumors except for *melanoma* (skin cancer) and *lymphoma* (lymph cancer).
    - *Carcinoma* refers to epithelial, or lining (outside), tissue.
    - *Sarcoma* refers to mesothelial (middle) tissue, such as connective tissue and muscle.
    - A skeletal muscle tumor is a *rhabdomyoma* if benign or a *rhabdomyosarcoma* if malignant.
    - An adrenal gland tumor is an *adenoma* if benign or an *adenocarcinoma* if malignant.

**Carcinogenesis**

- Chemical carcinogens (cancer generators) include vinyl chloride, insecticides, and cigarette smoke. These chemicals form adducts, which prevent certain genes from being read correctly.
- Biologic carcinogens include one known bacterium (*Helicobacter pylori*), which causes rapid replication of repair cells in the stomach as a response to inflammation; DNA viruses (hepatitis B, human papilloma virus), which replicates rapidly, and their replication gene may get incorporated into normal cells, making them malignant; and RNA viruses, such as HIV, which carry hitch-hiking cancer genes to normal cells.

- Physical carcinogens include radiation and UVB light (which break DNA strands) and asbestos (which "spears" cells, mixing cellular fluids and causing mutations).

**Mechanisms of carcinogenesis**

- Proto-oncogenes control normal cell division, which happens only 30 to 50 times per cell life.

- Damage to proto-oncogenes can disrupt the cell division signals, converting them to oncogenes that will then stimulate uncontrolled cell division.

- Tumor suppressor genes, such as P53, check genes during cell division and halt replication until the error is corrected. If the error cannot be corrected, the cell destroys itself. Damage to tumor suppressor genes can suppress these functions.

- Replication errors are related to the speed of replication; higher speeds create more errors. The high speed of cancerous cell division creates its own mutations.

- During promotion, the initiation process is reinforced by further cell division.

- During progression, cancerous cells invade other parts of the body, whether near (local invasion) or far (metastasis).

- A typical time from initiation to progression is 10 to 40 years. The fastest cancers known were leukemia cases after the nuclear explosion at Hiroshima, which took 7 years to present.

- During conversion, damaged cells permanently become cancerous.

- Cancer cells have several ways to move:

    - They can secrete chemicals that break down basement membranes.

- They can pull themselves along through the body.
- They can travel through blood vessels until they reach a capillary bed.

**Tumor angiogenesis**

- Tumors cannot grow to more than 2–3 mm in size without a blood supply.
- Cancerous cells can stimulate blood vessels to branch out to them; preventing this from happening is a fertile area of cancer research.
- Metastatic cancer is rarely curable, but there are three treatments:
    - Surgery to remove a tumor can leave cancerous cells behind, and the tumor may be inoperable.
    - Chemotherapy uses chemical poisons that travel to every cell in the body to kill cancerous cells.
    - Radiation can kill cancerous and noncancerous cells alike, but it needs a specific target.
    - More than 50% of all cancer patients can be cured, but this rate varies widely for a given type of cancer. ■

## Questions to Consider

1. Why is the word *potential* important in defining the characteristics of invasion and metastases in cancerous tissue?

2. What is the difference between primary cancers and metastatic cancer?

# Bibliography

**Essential Reading:**

Alberts, B., et al. *Molecular Biology of the Cell.* 4th ed. New York: Garland Publishing, 2002. An extensive and complete technical book for more detail on molecular biology than we could fit into this course.

Berry, A and James D. Watson. *DNA: The Secret of Life.* New York: Knopf, 2003. The story of DNA from the man who gave its structure to the world.

Cotran, Ramzi S., et al. *Robbins' Pathologic Basis of Disease.* 6th ed. W.B. Saunders, 1999. The leading pathology textbook for medical students.

Dorland, N. W., and P. D. Novak. *Dorland's Pocket Medical Dictionary.* 26th ed. Philadelphia: W.B. Saunders, 2001. Comprehensive dictionary of medical terms; very useful for quick reference.

Netter, Frank, and J. T. Hansen. *Atlas of Human Anatomy.* 3rd ed. Novartis Medical Education, 2003. Unquestionably the finest illustrations of anatomy by a great artist and surgeon. The staple for medical students for decades.

Tortora, G. J., and S. R. Grabowski. *Principles of Anatomy and Physiology.* 10th ed. New York: John Wiley & Sons, 2002. A fine combination of anatomy and physiology with excellent illustrations used in this lecture series. This should be the primary source of reading, extra information, and exam material.

**Supplementary Reading:**

Blaser, M. J. "The Bacteria behind Ulcers." *Scientific American* (February 1996), pp. 104–107. An excellent article about the current thinking on ulcer disease.

"Life, Death, and the Immune System." *Scientific American* (September 1993). Excellent review of the details of immunology.

Penrose, Roger. *The Emperor's New Mind: Concerning Computers, Minds and the Laws of Physics*. Oxford, NY: Oxford University Press, 2002. A classic on the working of the human brain.

Sacks, Oliver. *The Man Who Mistook His Wife for a Hat and Other Clinical Tales*. New York: Summit Books, 1985. Another classic from one of the great writers in neurology.

Talbot, Michael. *The Holographic Universe*. New York: Harper Collins, 1991. An in-depth work about the concept of holography.

Weinberg, Robert A. *One Renegade Cell: How Cancer Begins*. New York: Basic Books, 2000. A general introduction to cancer by a leading researcher in the field.

**Internet Resources:**

*Mayo Clinic*. More searchable information on many medical topics. http://www.mayoclinic.com/

*National Library of Medicine/National Institutes of Health*. Huge data bank of medical and scientific information. Free access. http://www.ncbi.nlm.nih.gov/

*New England Journal of Medicine*. Much information can be obtained without formal subscription. http://www.nejm.org

*Scientific American Medicine*. Frequently updated source of medical material. http://www.samed.com

*WebMD*. More information on specific diseases. http://www.webmd.com/

# Notes

# Notes

# Notes

# Notes

# Notes